Kew Gardens

Or, A Popular Guide to the Royal Botanic Gardens of Kew

William Jackson Hooker

Cambridge University Press

CAMBRIDGE
UNIVERSITY PRESS

University Printing House, Cambridge, CB2 8BS, United Kingdom

Published in the United States of America by Cambridge University Press, New York

Cambridge University Press is part of the University of Cambridge.
It furthers the University's mission by disseminating knowledge in the pursuit of education, learning and research at the highest international levels of excellence.

www.cambridge.org
Information on this title: www.cambridge.org/9781108065450

This edition first published 1858
This digitally printed version 2013

ISBN 978-1-108-06545-0 Paperback

CAMBRIDGE LIBRARY COLLECTION

Books of enduring scholarly value

Botany and Horticulture

Until the nineteenth century, the investigation of natural phenomena, plants and animals was considered either the preserve of elite scholars or a pastime for the leisured upper classes. As increasing academic rigour and systematisation was brought to the study of 'natural history', its subdisciplines were adopted into university curricula, and learned societies (such as the Royal Horticultural Society, founded in 1804) were established to support research in these areas. A related development was strong enthusiasm for exotic garden plants, which resulted in plant collecting expeditions to every corner of the globe, sometimes with tragic consequences. This series includes accounts of some of those expeditions, detailed reference works on the flora of different regions, and practical advice for amateur and professional gardeners.

Kew Gardens

The eminent British botanist Sir William Jackson Hooker (1785–1865) expanded and developed the Royal Botanic Gardens at Kew into a world-leading centre of research and conservation. Appointed its first full-time director in 1841, Hooker came to Kew following a highly successful period in the chair of botany at Glasgow University. He quickly began to extend the gardens, arranging for the building of the now famous Palm House and establishing the Museum of Economic Botany. This volume reissues Hooker's popular guides to the gardens (sixteenth edition) and to the museum (third edition), both published in 1858. Illustrated throughout, these documents reveal the areas and specimens accessible to a receptive Victorian public. Hooker's ten volumes of *Icones Plantarum* (1837–54) have also been reissued in this series, along with many works by his son and equally accomplished successor, Sir Joseph Dalton Hooker (1817–1911).

Cambridge University Press has long been a pioneer in the reissuing of out-of-print titles from its own backlist, producing digital reprints of books that are still sought after by scholars and students but could not be reprinted economically using traditional technology. The Cambridge Library Collection extends this activity to a wider range of books which are still of importance to researchers and professionals, either for the source material they contain, or as landmarks in the history of their academic discipline.

Drawing from the world-renowned collections in the Cambridge University Library and other partner libraries, and guided by the advice of experts in each subject area, Cambridge University Press is using state-of-the-art scanning machines in its own Printing House to capture the content of each book selected for inclusion. The files are processed to give a consistently clear, crisp image, and the books finished to the high quality standard for which the Press is recognised around the world. The latest print-on-demand technology ensures that the books will remain available indefinitely, and that orders for single or multiple copies can quickly be supplied.

The Cambridge Library Collection brings back to life books of enduring scholarly value (including out-of-copyright works originally issued by other publishers) across a wide range of disciplines in the humanities and social sciences and in science and technology.

Selected books of related interest, also reissued in the CAMBRIDGE LIBRARY COLLECTION

Amherst, Alicia: *A History of Gardening in England* (1895) [ISBN 9781108062084]

Anonymous: *The Book of Garden Management* (1871) [ISBN 9781108049399]

Blaikie, Thomas: *Diary of a Scotch Gardener at the French Court at the End of the Eighteenth Century* (1931) [ISBN 9781108055611]

Candolle, Alphonse de: *The Origin of Cultivated Plants* (1886) [ISBN 9781108038904]

Drewitt, Frederic Dawtrey: *The Romance of the Apothecaries' Garden at Chelsea* (1928) [ISBN 9781108015875]

Evelyn, John: *Sylva, Or, a Discourse of Forest Trees* (2 vols., fourth edition, 1908) [ISBN 9781108055284]

Farrer, Reginald John: *In a Yorkshire Garden* (1909) [ISBN 9781108037228]

Field, Henry: *Memoirs of the Botanic Garden at Chelsea* (1878) [ISBN 9781108037488]

Forsyth, William: *A Treatise on the Culture and Management of Fruit-Trees* (1802) [ISBN 9781108037471]

Haggard, H. Rider: *A Gardener's Year* (1905) [ISBN 9781108044455]

Hibberd, Shirley: *Rustic Adornments for Homes of Taste* (1856) [ISBN 9781108037174]

Hibberd, Shirley: *The Amateur's Flower Garden* (1871) [ISBN 9781108055345]

Hibberd, Shirley: *The Fern Garden* (1869) [ISBN 9781108037181]

Hibberd, Shirley: *The Rose Book* (1864) [ISBN 9781108045384]

Hogg, Robert: *The British Pomology* (1851) [ISBN 9781108039444]

Hogg, Robert: *The Fruit Manual* (1860) [ISBN 9781108039451]

Hooker, Joseph Dalton: *Kew Gardens* (1858) [ISBN 9781108065450]

Jackson, Benjamin Daydon: *Catalogue of Plants Cultivated in the Garden of John Gerard, in the Years 1596–1599* (1876) [ISBN 9781108037150]

Jekyll, Gertrude: *Home and Garden* (1900) [ISBN 9781108037204]

Jekyll, Gertrude: *Wood and Garden* (1899) [ISBN 9781108037198]

Johnson, George William: *A History of English Gardening, Chronological, Biographical, Literary, and Critical* (1829) [ISBN 9781108037136]

Knight, Thomas Andrew: *A Selection from the Physiological and Horticultural Papers Published in the Transactions of the Royal and Horticultural Societies* (1841) [ISBN 9781108037297]

Lindley, John: *The Theory of Horticulture* (1840) [ISBN 9781108037242]

Loudon, Jane: *Instructions in Gardening for Ladies* (1840) [ISBN 9781108055659]

Mollison, John: *The New Practical Window Gardener* (1877) [ISBN 9781108061704]

Paris, John Ayrton: *A Biographical Sketch of the Late William George Maton M.D.* (1838) [ISBN 9781108038157]

Paxton, Joseph, and Lindley, John: *Paxton's Flower Garden* (3 vols., 1850–3) [ISBN 9781108037280]

Repton, Humphry and Loudon, John Claudius: *The Landscape Gardening and Landscape Architecture of the Late Humphry Repton, Esq.* (1840) [ISBN 9781108066174]

Robinson, William: *The English Flower Garden* (1883) [ISBN 9781108037129]

Robinson, William: *The Subtropical Garden* (1871) [ISBN 9781108037112]

Robinson, William: *The Wild Garden* (1870) [ISBN 9781108037105]

Sedding, John D.: *Garden-Craft Old and New* (1891) [ISBN 9781108037143]

Veitch, James Herbert: *Hortus Veitchii* (1906) [ISBN 9781108037365]

Ward, Nathaniel: *On the Growth of Plants in Closely Glazed Cases* (1842) [ISBN 9781108061131]

For a complete list of titles in the Cambridge Library Collection please visit:
http://www.cambridge.org/features/CambridgeLibraryCollection/books.htm

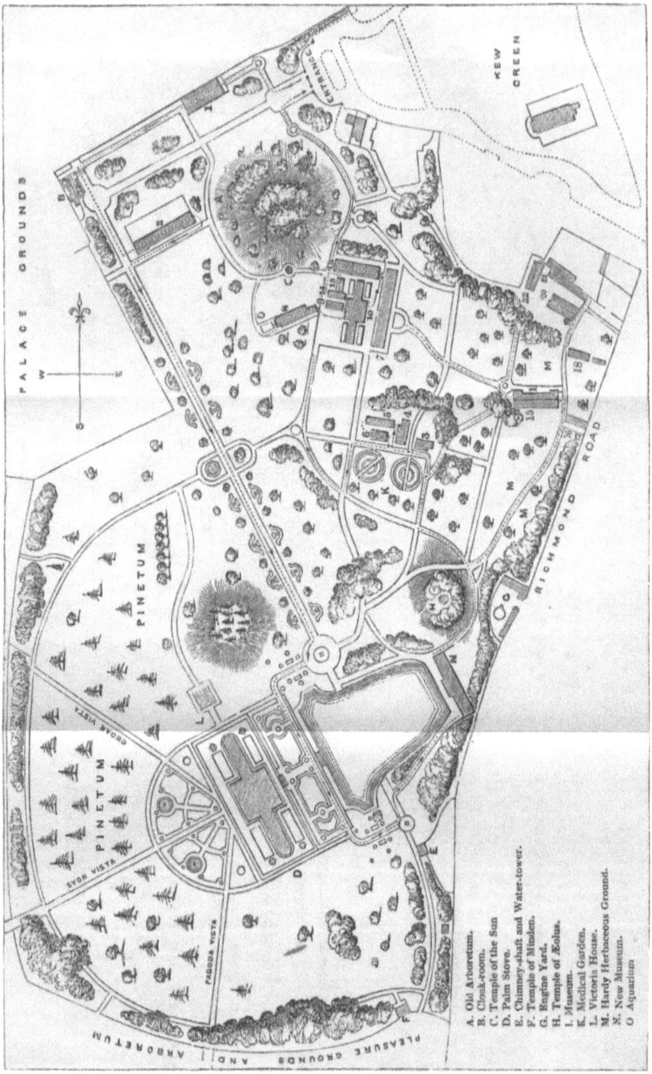

PLAN OF THE ROYAL BOTANIC GARDENS OF KEW.

The material originally positioned here is too large for reproduction in this reissue. A PDF can be downloaded from the web address given on page ivof this book, by clicking on 'Resources Available'.

KEW GARDENS

OR

A POPULAR GUIDE

TO THE

ROYAL BOTANIC GARDENS OF KEW.

BY

SIR W. J. HOOKER

K.H. D.C.L. F.R.A. & L.S.

CORRESPONDENT OF THE INSTITUTE OF FRANCE, ETC. ETC.

Director.

"Soft roll your incense, *Herbs* and *Fruits* and *Flowers*,
In mingled clouds, to Him whose sun exalts,
Whose breath perfumes you, and whose pencil paints."

SIXTEENTH EDITION.

LONDON

LONGMAN, BROWN, GREEN, LONGMANS, AND ROBERTS,

1858

" So sits, enthroned in vegetable pride,
Imperial Kew, by Thames's glittering side:
Obedient sails from realms unfurrow'd bring
For her the unnamed progeny of Spring.

" Delighted Thames through tropic umbrage glides,
And, flowers antarctic bending o'er his tides,
Drinks the new tints, the sweets unknown inhales,
And calls the sons of science to his vales.
In one bright point admiring Nature eyes
The fruits and foliage of discordant skies,
Twines the gay flow'ret with the fragrant bough,
And binds the wreath round George's regal brow.

" Sometimes, retiring from the public weal,
One tranquil hour the Royal Partners steal,
Through glades exotic pass, with step sublime,
Or mark the growth of Britain's happier clime."

LONDON:
Printed by SPOTTISWOODE & Co.
New-street Square.

PREFACE.

A Guide, which should indicate to strangers the more remarkable features in the Royal Botanic Gardens of Kew, and point out some of the many interesting plants cultivated there, has long been a desideratum. Of late, this want has been peculiarly felt, because of the great extent of ground, the number of plant-houses, and the amazing increase of the collection. The very fact, however, of the continued additions to the plants, combined with their rapid growth under good cultivation, renders any approach to a *perfect* Guide or Handbook a very difficult, if not an impossible, task; for, though it is true that only a reference to the more remarkable objects can be desired, yet the frequent arrival of novelties must, of necessity, cause such a book to become, in a measure, imperfect soon after its publication. This circumstance, along with the constantly increasing bulk, entails the incessant removal of plants from one house to another; thus the individuals that are recorded as occupying one particular greenhouse or stove may require shifting the very next day. It is eminently needful to warn our readers of this circumstance, because they will thus comprehend how it happens that a plant stated to exist in Plant-house No. 1., for instance, may not at the time of their visit be found there. To a certain extent, and owing to the causes just mentioned, individual specimens cannot be stationary for a great space of time; still we endeavour to retain them in the places indicated as long as possible; and, if a large plant of peculiar interest be necessarily removed, we shall, as often as we can, replace it with a smaller individual of the same kind. When this is impracticable, and any particular plant is not seen where the Guide-Book states it to be, the *Index* will probably refer to the page where it is noticed.

The beauty of these grounds and of the plants which they contain, combined with the liberal admission granted by Government, attracts, as may be supposed, great crowds of visitors; and a few needful regulations, over and above those expressly posted in the grounds, may be here appropriately given.

1. Smoking, or eating and drinking, or the carrying of provisions of any kind into the Gardens, are strictly forbidden. No dogs can be admitted.

2. No packages or parcels, bags or baskets, are allowed to be carried within the grounds. All such articles must be deposited at the gate of entrance while the owners make the tour of the Gardens.

3. No person attired otherwise than respectably can enter, nor children too young to take care of themselves, unless a parent or suitable guardian be with them: the police have strict orders to remove such, as also persons guilty of any kind of impropriety. Nor can large schools have admission, except in accordance with the printed regulations to be seen at the gate.

4. It is by no means forbidden to walk upon the lawns; still it is requested that preference be given to the gravel-paths, and especially that the lawn-edges parallel to the walks be not made a kind of footway, for nothing renders them more unsightly. It might scarcely be thought needful to say, that all play, leaping over the beds, and running, particularly on the mounds and slopes, are prohibited; yet the latter has been practised, and so heedlessly, that very serious injuries have resulted from falls, and grievously scarred faces have been the memento of such folly. The Gardens are intended for agreeable recreation and instruction, not for idle sports.

5. It is requested that visitors will abstain from touching the plants and flowers: a contrary practice can only lead to the suspicion, perhaps unfounded, that their object is to abstract a flower or a cutting, which, when detected, must be followed by disgraceful expulsion.

6. It is particularly requested that visitors will enter the Plant-houses by the doors indicated for the purpose; if they do otherwise, and come in by opposite ones indiscriminately, they will meet and pass each other, which the narrowness of the walks renders difficult; and this must occasion inconvenience to all parties, and often injury to the plants.

The accompanying Plan of the Gardens and Plant-houses will, it is expected, prove useful; and a stranger to the ground and the collection may do well to follow the route indicated by *dotted lines*, as the most convenient for giving a tolerably complete survey of the whole.

7. No children's chaises are admitted; but *grown-up invalids* can have permission to visit the Gardens in wheel-chairs, on a written application to the Director, giving name and residence.

More might be said on these heads; but the Director, while bearing willing testimony to the excellent conduct of the many thousands who frequent the Gardens, prefers to rely on the good sense and honourable feelings of the visitors, and the value they must attach to the privileges here afforded, rather than multiply restrictions which may not be absolutely required.

The *Botanic Gardens* are open every week-day from one till sunset: and no person can be admitted at other hours *except on business*. On Sundays they are open from two to seven, or sunset, in winter.

N.B. The *Royal Pleasure-Grounds* or *Arboretum*, sometimes by strangers confounded with the Botanic Gardens, constitute a separate though adjoining portion of ornamental ground, accessible daily from May to Michaelmas, by three gates,—two in the road leading from Kew to Richmond, called the *Lion* or *Pagoda Gate* and the *Unicorn Gate*, and one by the river-side, nearly opposite Brentford Ferry, called the *Brentford Gate*,—besides the gates of communication to and from the Botanic Gardens.

To strangers desirous of presenting Plants or Museum objects from abroad to Kew, we here take leave to mention, that, in despatching packages and parcels, the quickest mode of transit is always the best. When sent by the Queen's ships or the Royal Mail, or the Peninsular and Oriental Company's steamers, the address should be:—

"To the SECRETARY of the ADMIRALTY,
"LONDON.
"For Sir Wm. J. Hooker,
"Royal Gardens, Kew."

If by merchant or other vessels, the direction is simply,

"To Sir WM. J. HOOKER,
"Royal Gardens, Kew,
"LONDON."

KEW GARDENS.

BRIEF HISTORIC NOTICE.

It is generally known that considerable changes in the Royal Botanic Grounds of Kew were contemplated about the year 1840, when, from being a private garden belonging to the Royal Family, and maintained by funds from the Board of Green Cloth, it was liberally relinquished by her present Majesty, Queen Victoria, and placed under the control of the Commissioners of Her Majesty's Woods and Forests, with the view of rendering it available for the general good. The public, having since been freely admitted to the Gardens under a few needful regulations, must have observed the many alterations and improvements effected under the sanction of the above-mentioned Commission, and cannot fail to desire some information respecting them. It is with a view to satisfy such laudable curiosity, and to increase the interest with which the Gardens are visited, that this Guide is now compiled.

We shall not here enter into the full and early history of the Royal Gardens of Kew: a few statements are, however, necessary, and we have selected them from the best authorities.

About the middle of the seventeenth century, the spot that now forms the Royal Gardens of Kew, together with a residence called *Kew House,* belonged to R. Bennett, Esq., whose daughter and heiress married Lord Capel. There is a handsome white marble monument to this lady in Kew Church. Kew House and Grounds then passed into the hands of Mr. Molyneux, who was secretary to King George II. (when Prince of Wales), and who married Lady Elizabeth Capel. He was well known as a man of literature and an astronomer. With an instrument of Mr. Molyneux's own construction, and in those very grounds, Dr. Bradley made the valuable discoveries relating to the fixed stars, to commemorate which an inscription was placed by the late King William IV. on the pedestal of a sun-dial, which stands on the identical spot which had been occupied by Dr. Bradley's telescope, upon the lawn, opposite to the present palace.

The Prince of Wales, who was son to George II., and father to George III., admiring the situation of *Kew House*, took a long lease of it from the Capel family about the year 1730, and began to form the pleasure-grounds, then containing about 270 acres. They were completed by his widow, Augusta, Princess Dowager of Wales, who delighted in superintending the improvements, then conducted upon a most extensive scale. At this time Sir W. Chambers was employed in decorating the Gardens at Kew with temples, &c., an account of which he published in a large folio work with many plates, (dedicated to the Princess Dowager of Wales,) under the title of "Plans, Elevations, Sections, and Perspective Views of the Gardens and Buildings at Kew, in Surrey, the Seat of H. R. H. the Princess Dowager of Wales."

The Exotic Department of this Garden was commenced by the same Princess, and much favoured by the Earl of Bute, about the middle of the eighteenth century. Many of the finest foreign trees were contributed by Archibald Duke of Argyle (styled by Horace Walpole the Tree-monger), who sent them from his once richly-stored garden at Whitton, near Hounslow.

We find that in the year 1759, Mr. W. Aiton, a pupil of the celebrated Philip Miller, of the Chelsea Physic-Garden, was placed in charge of the Botanical Gardens at Kew,—a gentleman no less distinguished by his private virtues than his knowledge of plants, and great skill in cultivating them. His professional abilities quickly procured him the notice of the late Sir Joseph Banks, and a friendship commenced which subsisted between them for life.

About the year 1789 His Majesty George III. purchased Kew House, which was soon afterwards pulled down, and its furniture removed to an older mansion, since known by the name of *Kew Palace*, and once the property of Sir Hugh Portman, who is mentioned as "the rich gentleman who was knighted by Queen Elizabeth at Kew." This small but picturesque red brick dwelling, which appears to be of the date of King James, or Charles I., was purchased in 1781 for Queen Charlotte (who died there); and it was long the favourite suburban residence of the Royal Family. Her Majesty evinced much interest in the increase of the collection of plants; and justly does the late Sir James E. Smith, President of the Linnæan Society, bear testimony to the Queen's love of botany, when he says "that the genus *Strelitzia* (so called by Mr. Aiton) stands on the sure basis of botanical knowledge and zeal, few persons having cherished the study of nature more ardently, or cultivated it so deeply, as Her Majesty." Under such auspices, and aided by the enlightened patronage of Sir Joseph Banks, it was only to be expected that the Gardens of Kew should become celebrated all over the world. So early as 1760, the great or old Stove (No. 8. of the Plan), 114 feet long, was built by Sir William Chambers.

In 1761 the noble Orangery (No. 2. of the Plan) was erected also by Sir William Chambers. It measures 145 feet in length, its width is 30 feet, and its height 25 feet. In the same year was added the very elegant Temple of the Sun, as it is called, of the Corinthian order (C. of the Plan); and some young trees were planted near, which are now grown to be among the most beautiful in the Gardens, particularly an *Oriental Plane* and a *Turkey Oak.* Such had been the increase of plants, that, in the year 1788, a greenhouse was built for Cape plants (since demolished), 110 feet long; and another for the vegetable productions of New Holland, nearly the same size (No. 10. of the Plan), was added in 1792. (This latter has been much enlarged and improved under the name of the "Australian House.")

A catalogue of the plants in the Exotic Garden of Kew was published by Dr. Hill in 1768, and a second edition the following year.

A far more elaborate and important work appeared in 3 vols. 8vo., accompanied by some admirable plates, the *Hortus Kewensis* of William Aiton, in 1789, giving an account of the several foreign plants which had been introduced into the English gardens at different times, amounting to 5,600 in number; and so much was it esteemed that the whole impression was sold off within two years. Mr. Aiton did not long survive this publication, for he died in 1793, in the sixty-third year of his age, and lies buried in the churchyard at Kew, near the graves of his distinguished friends, Zoffany, Meyer, and Gainsborough. He was succeeded by his son, W. Townsend Aiton, Esq., who was no less esteemed by King George III. than his father had been, and who, besides conducting the botanical department, and taking charge of the extensive pleasure-grounds, was also employed in the improvement of the other Royal gardens, in all which he displayed great skill and judgment, and an intimate acquaintance with his profession.

The voyage of Captain Cook and Sir Joseph Banks round the world; those of Captain Flinders and Mr. Robert Brown (*Botanicorum Princeps*), and of Mr. Allan Cunningham, to Australia; the expeditions of Bowie and Masson respectively to Brazil and the Cape of Good Hope—all these enriched the Gardens of Kew with the vegetable productions of the southern hemisphere, to an extent unparalleled before or since: besides which, other collectors were employed abroad during a long period of years in various countries; and the produce of their researches was deposited at Kew. On various occasions, especially during the life of King George III., other houses, stoves and pits were erected, as occasion required; but it must be confessed that, on the demise of that revered monarch and of Sir Joseph Banks, whom His Majesty so much delighted to honour, and who died shortly after the King, the establishment languished and suffered from want of Royal and scientific encouragement. During the reigns of George IV. and William IV., with the exception of a few plants being transmitted by occasionally employed collectors, and one hothouse being erected by the last-mentioned sovereign (and it is but right to add that this conservatory is eminently handsome and ornamental), the Botanic Gardens retrograded rather than flourished; and matters must have been much worse, but for the truly parental affection cherished towards it by Mr. Aiton, and the able exertions of his foreman (now the curator), Mr. John Smith. Throughout the country an opinion existed, which soon began to be loudly expressed, that either the Gardens should be entirely abolished or placed upon a very different footing, and rendered available, as a great popular yet scientific establishment, for the advantage of the public.

Government was, happily, ready to respond to this latter feeling; and in 1838, the Lords of Her Majesty's Treasury appointed a committee to inquire into the management, condition, &c. of the Royal Botanic Gardens. The result was, that in May, 1840, a return was made to the House of Commons, in the shape of a report by Dr. Lindley, who, at the desire of the committee, had surveyed the Gardens, in conjunction with two well-known practical gardeners.

Strangers, or persons not well acquainted with the vicinity of Kew, often entertain very incorrect notions of this establishment; nor can such be wondered at, seeing for how long a time it was the private garden of the Royal Family, and taking also into account its extensive and highly-varied nature. It may be interesting, especially as exhibiting most forcibly the change that has since taken place, to describe in few words the extent and condition of the grounds at the time of this investigation, namely, in 1840. They then consisted of—

1. *The Grounds immediately about the existing Palace of Kew*, which were of small circuit, lying near the river, and consisting mainly of those of the great edifice or Palace *, begun by Mr. Wyatt in the reign of His Majesty King George III., and soon afterwards pulled down, and the grounds of the present Palace. The boundary

* The site of this Palace is now, with the sanction of Her Majesty, used as a nursery for the supply of the London Parks.

is the river on the north side, the Pleasure-Grounds on the south and west, and the Botanic Garden on the east.

2. *The Botanic Garden proper,* which contained at the time in question 11 acres, or thereabouts, of very irregular outline; bounded on the north partly by the gardens of those residences, mainly Crown property, which stand on the south side of Kew Green, in part by the Green itself, from which it was separated by a handsome railing, and in part by the gardens of His late Majesty the King of Hanover; westward, by the grounds of the Palace above-mentioned; eastward, by what were then the Royal Kitchen and Forcing-Gardens (now a part of the Botanic Garden); and south by the Pleasure-Ground.

3. *The Royal Kitchen and Forcing-Gardens,* situated between the Botanic Garden and the Richmond road, comprising about fourteen acres. (This portion has been, as just observed, added to the Botanic Garden.)

4. *The Pleasure-Ground* or *Arboretum,* comprising 270 acres of wood, shrubbery and lawn, lying to the south of the Botanic Garden, and bounded by the Richmond road on one side, and the river on the other. For some years this extensive and beautiful area had been thrown open only twice a week during the summer (now daily during that season, and the public are admitted at three different entrances).

5. South of this, and stretching between the Richmond road and the Thames, almost into the lower part of Richmond, lies *Richmond Old Park,* or the *Old Royal Deer-Park,* as it is sometimes called; a noble extent of pasture, comprising about 400 acres, interspersed with many fine trees; distinguished by the *Observatory* erected by George III., and now liberally granted to the use of the British Association, where that scientific body has carried on an interesting series of experiments on terrestrial magnetism.

The report of Dr. Lindley, mentioned above, has reference only to the second of these divisions, namely, the *Royal Botanic Gardens,* which are stated to "include many fine exotic trees and shrubs, a small collection of herbaceous plants, and numerous specimens of grasses." Ten different stoves and greenhouses then existed; most of which have been either condemned and pulled down as unworthy of the Gardens, or so greatly altered as to be no longer recognizable under Dr. Lindley's description.

It resulted from this investigation, that the whole of the Gardens, Pleasure-Grounds and Park was transferred to the department of the Commissioners of Her Majesty's Woods and Forests. Mr. Aiton, on the eve of the fiftieth anniversary of his holding office, retired from the charge of the Botanic Gardens; and the present Director received instructions from the Board to enter upon his important duties in the spring of the year 1841, and to prepare, as speedily as possible, a Report of those alterations which were deemed essential for rendering the Gardens useful to the public at home and to our colonies abroad. Many useful suggestions on these heads were offered by Dr. Lindley in the before-mentioned document, especially the following: — " A national garden ought to be the centre, round which all minor establishments of the same nature should be arranged: they should be all under the control of the chief of that garden, acting in concert with him, and through him with one another, reporting constantly their proceedings, explaining their wants, receiving their supplies, and aiding the mother-country in every thing that is useful in the vegetable kingdom. Medicine, commerce, agriculture, horticulture, and many valuable branches of manufacture, would derive much benefit from the adoption of such a system. From a garden of this kind, government would be able to obtain authentic and official information on points connected with the founding of new colonies: it would afford the plants there required, without its being necessary, as now, to apply to the officers of private establishments for advice and assistance."

Changes of a highly important character could not fail in suggesting themselves to the Director, on his becoming intimately acquainted with the minutiæ of the establishment, many of which it were tedious to narrate in this place.

One of the first was to open the Botanic Gardens for the admittance of the public daily. Not only the Grounds but the Plant-houses and Museum are open to visitors; the number of whom, it is needless to say, is very considerable*; yet, what is peculiarly gratifying, and contrary to the anticipation of many persons, this privilege has been rarely abused. In the few cases of an opposite line of conduct, the consequent detection (which must be expected where trustworthy men are necessarily dispersed through the Gardens at their various occupations) has proved its own punishment.

Next to the facility and consequent pleasure and instruction to the public, the enlargement of the ground was an important object. The limit of the Garden was not, indeed, exactly defined where it met the precincts of the residence of His Majesty the King of Hanover; but permission was soon obtained to include within the Botanic Garden

* It may not be uninteresting to our readers to state the gradual increase of visitors, since the Botanic Gardens were thus daily thrown open to the public. The amount of visitors

in 1841 was	9,174	in 1850 was	179,627
1842 „	11,400	1851 „	238,900
1843 „	13,492	1852 „	231,210
1844 „	15,114	1853 „	331,210
1845 „	28,139	1854 „	339,164
1846 „	46,573	1855 „	318,818
1847 „	64,282	1856 „	344,140
1848 „	91,708	1857 „	361,978
1849 „	37,865		

all the ground immediately about the Conservatory and Orangery, which greatly enhanced the beauty of the view, and added between 3 and 4 acres. This augmentation to the limits, however, was, from its small extent, rather to be considered ornamental than useful. Application was made by the Chief Commissioner of Woods and Forests, to the Queen, for a grant of land from the contiguous Pleasure-Ground, which might afford the means of forming a *Pinetum* (or a collection of plants of the Pine-tribe) suited to such an establishment, and also of erecting a Palm-Stove, or tropical house, equally worthy of the place and the nation. Her Majesty was graciously pleased to assent to this request; and a portion of the Pleasure-Ground, comprising about 47 acres, and including a piece of water, was surveyed, and permitted to be enclosed within a light wire fence, which still gives to view the rest of the Pleasure-Ground, and adds to the beauty of the Botanic Gardens, which, thus augmented, contained 60 acres.

Again, in the winter of 1846-7, orders were received for abolishing the Royal Kitchen and Forcing Gardens of Kew, as such, and incorporating them with the Botanic Grounds, which has already been done, thus adding 15 more acres to the scientific portion of the grounds (75 acres in all).

But changes now come to be noticed that have been effected within the above-mentioned *Botanic Garden* grounds; for, in the same ratio that hardy plants required more space, so did the tender plants need increased accommodation; and plans were accordingly given in for those improvements, by which such a transformation is effected in the aspect of the place, that persons who have not visited Kew Gardens for a few years can scarcely recognise the localities. We shall describe, with all possible brevity, the present condition of the Royal Botanic Gardens, and at the same time indicate the objects most worth the attention of a stranger, both in the open ground and in the several plant-houses. There is a separate Guide-Book to the extensive and valuable contents of the two Museums of Economic Botany within the Gardens.

GUIDE TO THE BOTANIC GARDENS.

On approaching the Botanic Gardens by the new entrance at the head of Kew Green, the visitor cannot fail to be struck with the beauty of the richly ornamented gateway, erected in 1845–6, by the late Mr. Walker, of York, from a design of Decimus Burton, Esq. Passing through it, the main walk takes a westerly course, and, besides catching a distant view of Kew Palace, attention will be attracted on the left by the fine trees of the OLD ARBORETUM, a collection of hardy exotic trees and shrubs. On the lawn on the right-hand side of this walk, among other recently planted and recently introduced trees, is the graceful *Cryptomeria Japonica*, a plant of the Pine-kind, native of Northern China and Japan, with other young evergreen trees or shrubs, and a Palm of Northern China, *Chamærops excelsa*, which has been found to bear our climate for some years past with little or no protection; and on the other side of the walk, on the outskirts of the old Arboretum, will be seen good specimens of *Douglas's Pine* (*Abies Douglasii*), *Pinus Sabiniana*, &c. The Plant-house, which here comes directly in view, is

No. 1. THE CONSERVATORY;

A handsome stone building, of classical design, sometimes called the Architectural Green-house. This fine structure was removed hither, by order of His Majesty William IV., from Buckingham Palace in 1836. It is one of the three* Conservatories that had been erected in the gardens there, heated by innumerable coils of small pipes, fixed by Mr. Perkins, and is now filled with an extremely rich collection of Australian trees and shrubs, chiefly *Myrtaceæ*, *Leguminosæ*, *Proteaceæ;* the latter a family of plants, so named in consequence of the very varied character of the stems, leaves, and inflorescence, yet agreeing in the essential character of the flowers and fruit. They are handsome evergreen shrubs, or small trees, constituting much of the so-called "*Scrub*" of New Holland. Among the numerous kinds of this extensive group the *Banksias* and *Dryandras* are the most remarkable, and the handsome *Waratah*

* The second is still a Conservatory at Buckingham Palace, while the third has been there converted into a Royal Chapel.

(*Telopea speciosissima*, fig. 1.) Their foliage, though harsh and rigid, has something of the Fern character; and the flowers, especially of the *Banksias*, are arranged in bunches or tufts resembling a bottle-brush. It is in the winter and spring season generally that they are in flower. The smaller and younger plants in this house are raised chiefly from Swan River seeds, sent by Mr. Drummond, and are of great rarity and value. Here, too, is placed the curious *Hand Plant* (*Cheirostemon platanoides*, fig. 2.) of Mexico, with leaves resembling those of a Plane-tree; the stamens, resembling the fingers of the human hand, probably recommended this curious plant as an object of worship. At the period of Humboldt's visit, the only tree then known in Mexico was held sacred.

1

TELOPEA SPECIOSISSIMA.

2
THE HAND-PLANT.

We are supposed to have entered this conservatory by the eastern door: on quitting it at the west end, the path leads towards the Palace, with a vista in front; of which the view extends past the front of the Palace and across the river to the grounds of Syon House, the mansion of His Grace the Duke of Northumberland. The main walk soon takes a southerly direction a little before coming to a Cloakroom, where ladies will always find a place of rest or shelter in wet weather, and where their umbrellas or cloaks can be deposited by those who contemplate a long walk, under the care of an obliging female attendant. Here, on turning to the left, the visitor enters upon the grand and favourite promenade of the Garden. Proceeding, the attention is drawn by a large edifice on the left facing the south, which we still call by its original name,

No. 2. THE ORANGERY;

Which is used to shelter, in the winter, numerous large and half-hardy trees and shrubs, especially tender *Pines*, many of which are of great rarity and value. The house was erected by Sir William Chambers in 1761*, and it bears on the front, in two

* Not 1751, as incorrectly inscribed on the shields in the facade.

shields, the initials of Augusta, Princess Dowager of Wales, who, as already mentioned, took a great interest in the Gardens of Kew, and to whom Sir William Chambers dedicated, in 1763, his "Designs of Her Royal Highness's magnificent Villa at Kew." The two ends of this edifice were altered and furnished with large windows in 1842, and they bear the royal arms and that date accordingly. It was originally destined for, and filled with, orange trees, till 1841, when they were removed to Kensington Palace (with the exception of a few), and their places supplied by a very miscellaneous collection of trees and shrubs, which had become too large for the other greenhouses. The tenderer *Pines* (*Coniferæ*) constitute, perhaps, the most prominent feature in this house, when it has received its inmates for the autumn and winter. Here, at the latter seasons, may be seen the noblest specimens in Europe of the *Norfolk-Island Pine* (*Araucaria excelsa*), remarkable for their beautifully drooping and graceful branches, which almost vie with ostrich plumes;—the *Araucaria columnaris* of New Caledonia, where its stately column-like appearance was noted by the circumnavigator, Captain Cook, under the name of *Cupressus columnaris;* the *Pencil Cedar,* as it is commonly called,—no cedar, indeed, but an American *Juniper* (*Juniperus Bermudiana*);—the *Moreton-Bay Pine* (*A. Cunninghami*), together with another species from Moreton Bay, N. E. Australia, resembling, in its foliage, the Chili Pine (*A. imbricata*), and long the only specimen of the tree in Europe. It was discovered in the high lands, near Moreton Bay, by the late J. G. Bidwill, Esq., and having been by him presented to the Gardens, it justly bears his name (*A. Bidwilli*): its full-grown cones are as large as a child's head; and, as the seeds of the Chili Pine are eaten in South America, so are these eagerly sought for, as an article of food, by the aborigines of Australia, who at the proper season migrate to the pine-woods for the sole purpose of collecting them. The *Brazilian Pine* (*A. Braziliana*), the *China broad-leaved Pine* (*Cunninghamia lanceolata,* fig. 3.), graceful *Pines* from the Himalaya Mountains, and several others, equally rare, from Mexico and elsewhere, are here; all needing protection during the winter.

8

CUNNINGHAMIA LANCEOLATA.

In this house some large *Gum-Trees* of Australia (*Eucalyptus*) are easily recognizable,—one kind is the rapidly-growing *Gum-Tree* described by Mr. Backhouse, when he says: "It is the most gigantic tree of Van Diemen's Land, and there called *Stringy Bark.*"

Some of the specimens exceed 200 feet, rising almost to the height of the Monument in London before branching: their trunks also will bear comparison with that stately column both for circumference and straightness. Here are unquestionably the finest specimens in Europe of the famous *New Zealand* or *Cowdie* (sometimes called *Cowrie*, or *Kauri*) *Pine* (*Dammara australis*), the gift (with many other rarities) of Admiral Sir William Symonds, R.N., the late Surveyor-General of the Navy; than whom no person was more competent to estimate its value for spars for the British navy. Ship-loads are imported to supply the Royal Dockyards. It affords also copiously a valuable gum-resin.

In the same house may be seen the *Camphor-Tree* (fig. 4.) of Japan (*Laurus Camphora* L.); but so miscellaneous is the collection here, and so variable, in consequence of the plants and shrubs being moved in summer to different parts of the lawns and walks, that it is unnecessary to enumerate any more of them. The largest, and the rarest, and the best, especially the Norfolk-Island Pines, will generally be found at that season placed in the vicinity of the Orangery.

4

CAMPHOR TREE.

The visitor, on quitting this building, will probably be disposed to return to the main path leading to the Palm-house and to the Victoria-House; and he can hardly fail to be struck with the beauty of this noble walk, and with the judgment shown by Mr. Nesfield in the disposition and shape of the beds of shrubs and flowers. Alternating with the large beds are planted two lines of *Deodars*, designed eventually to form an avenue of this stately and graceful tree. Secondary lines are composed of Junipers, Cypresses, and other allied plants. The Deodar line is in one place interrupted by a beautiful *Turkey Oak*, too fine a specimen to be sacrificed, even for the sake of perfecting the avenue: the whole vista terminating on the south by the very handsome tower, which, while it performs the duty of a chimney-shaft for the Palm-house, includes a great tank (that receives its supply of water through the means of a steam-engine), by means of which the ornamental water and its jet are supplied, and applicable to the watering of the plants of the whole garden and all the houses. Westward on the lawn may be seen a fine group of *Elms*, known by the name of "the Seven Sisters" (so called in allusion to the daughters of His Majesty George III.), and two noble *Limes*; and in this direction walks have been recently made (see the Plan) to branch off and conduct through the newly-planted *Pinetum* (collection of *Coniferæ* or *Pines*); by which the visitor may approach the Victoria-House or the western entrance of the Palm-house by the Syon vista, or extend his walk to the tower and ornamental water.

Returning, however, to the principal promenade, and continuing south, we find a handsome piece of water with a jet or small fountain, enlivened by swans, and by other aquatic birds from the Zoological Society of London. The *Canadian Canoe,* formerly moored here in summer (made of the *Paper Birch, Betula papyrifera*), became leaky and decayed. A model, made of the same material, is in the New Museum. A branch-walk to the right brings us to the recently finished

TROPICAL AQUARIUM.

(L on the Plan), with its circular tank, 36 ft. in diameter, originally intended for the cultivation of the splendid plant which bears the name of our most gracious Queen, *Victoria regia ;* an inhabitant of the still waters or Igaripés of tropical America, especially in Guiana, and the tributaries of the Amazon. Our gardens are indebted for the possession of it to Dr. Rodie, and to Mr. Luckie, who sent fresh seeds in phials of water: — "It was on the 1st of January, 1837," writes Sir Robt. H. Schomburgk, the eminent traveller and recent discoverer of this extraordinary vegetable production, "while ascending the river Berbice, that some object attracted my attention which I could not comprehend ; but animating my crew to increase the rate of paddling, we soon came opposite a truly vegetable wonder. All calamities were forgotten. I was a botanist, and felt myself rewarded. Here were gigantic orbicular leaves, floating on the water, five and six feet in diameter, with a broad rim, light-green above and purple-crimson below ; while, in character with this wonderful foliage, I saw flowers a foot and a quarter (15 inches) across, fragrant, white, with a pink, at length deep rose-coloured, centre."* It was not till 1849, though many attempts had been made previously, that we succeeded in rearing plants from the seeds, which we gladly distributed among our most distinguished horticultural friends. In 1850 our plants came to perfection, and have ripened seed abundantly, so that we shall rarely be without flowering specimens in the summer. In the winter the plant lies nearly dormant.

Other water-plants occupy the different parts of the tank, and

* See, for a history of this remarkable aquatic, Botanical Magazine for 1847, tab. 4275—4278; and for a more full account, see "Figures and Description of the *Victoria Water-Lily*," published by Messrs. Reeve, in imperial folio. The plant itself being found to succeed best in the Plant-House, No. VI., it is transferred there.

amongst them the beautiful "Sacred Bean" of India (*Nelumbium speciosum*, fig. 5.), κυαμος of the ancients, the *Euryale ferox*, the

5

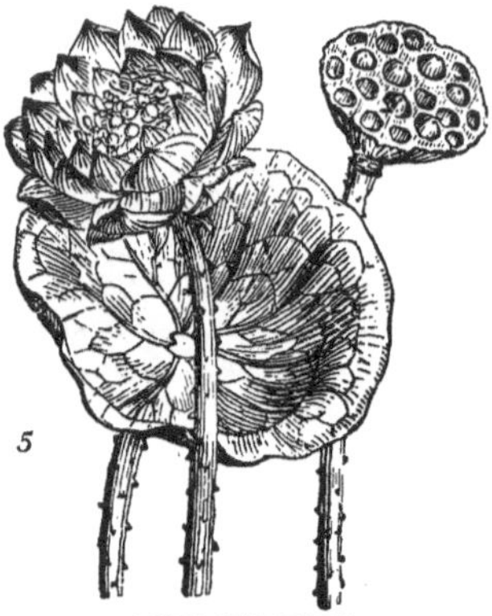

SACRED BEAN OF INDIA.

Papyrus, large leaved *aroideæ*, &c.

The adjacent

PALM-HOUSE, OR PALM-STOVE,

(Fig. 6.), completed in 1848, may be said to be the glory of the Gardens. It is built from the design of Decimus Burton, Esq.; and the iron-work is executed by Mr. Turner, of the Hammersmith Works, Dublin; the brick and stone work by Messrs. Grissell and Peto; and the boilers by Messrs. Burbidge and Healy: all working in concert with the Director and Curator of the establishment, who are responsible for the successful cultivation of the plants. As the public have the opportunity of inspecting this noble stove, we shall content ourselves with remarking, that the shell or external frame consists of a centre and two wings, occupying an area 362 feet in length; the centre is 100 feet wide and 66 feet in height to the summit of the lanthorn; the wings 50 feet wide and 30 feet high. The whole is of iron, stone, brick, and sheet-glass, the latter slightly tinged with green, at the suggestion of R. Hunt, Esq., of the Geological Survey, in order to temper the too powerful rays of light, which is thus in a measure accomplished. The extent of glass for covering this vast building is about 45,000 square feet. The ribs are inserted in enormous blocks of Cornish granite, placed on the most solid concrete. The central portion of the building (138 feet long and 100 feet wide) has a substantial gallery all round at the height of 30 feet from the floor, ascended and descended by light

PALM-STOVE.

spiral staircases, so as to give the opportunity of viewing the plants from above as well as below by bringing the spectator on a level with the summits of many of the loftiest, and also affording the means of watering the plants from above. The whole interior is heated by hot-water pipes and tanks (the hot-water pipes, $4\frac{1}{2}$ inches in diameter, are estimated to extend 24,000 feet in length, and the hot-water tanks 1000 feet), also constructed by Mr. Turner, judiciously distributed under the tables and beneath the level of the floor. To avoid the unsightliness of a chimney attached to, or even placed near, so noble a structure, the smoke is conveyed by underground flues, within a brick tunnel 7 feet high (from the underground furnaces, twelve in number), to a distance of 479 feet from the House; where a shaft or ornamental tower is erected, 96 feet in height, so situated and of such a form as to be an architectural object when seen from the main walk. Near the base of the tower, between it and the Richmond road, is the coal-yard, concealed by shrubs; and here too, within the underground tunnel above alluded to, is a railroad, for the purpose of conveying coals to the furnaces, and for bringing away the ashes.

If, before entering, we make the tour of the terrace of the Palm-house, we shall find that the immediate vicinity, at the east and west fronts, is laid out with ornamental parterres for flowers or shrubs. To the westward is a considerable area, or lawn, of some 25 acres, constituting a PINETUM, where are now seen all the *Coniferous plants* which bear the open air; while from the great western entrance of the Palm-house three vistas radiate at equal distances, commanding views through the Pleasure-Grounds. One, inclining to the south, in the direction of the Pagoda (best seen, perhaps, from the gallery of this Palm-house), is bounded on each side, for a length of 2800 feet, by a line of scarlet Thorns, alternating with spiry Evergreens, as Cypresses, Thujas, and Junipers, constituting a lawn avenue. Outer lines of Deodars are also planted, which will eventually form the permanent vista. The second, or Syon vista, completed in the spring of 1852, and looking west, extends to the river, nearly $\frac{3}{4}$ of a mile, has a broad gravel walk, and is bordered with Deodars alternating with Limes. The third, looks towards a fine Cedar in the direction of Brentford.

We must now direct attention to some of the numerous objects in the Palm-house, a structure especially intended for the cultivation of those "Princes" of the vegetable kingdom, but by no means wholly confined to them. The Palms constitute, however, a splendid and striking feature of its vast area, and are seen to most advantage from the gallery above. Among the *loftiest* Palms in the House are two kinds of Cocoa-nut, of which one (*Cocos plumosa*) is an old inhabitant of these Gardens, and the other (*Cocos coronata*) was presented by Sir George Staunton, Bart., having been transported, though of so vast size, by railway, from that gentleman's beautiful seat, Leigh Park, Hants, in a case 42 feet in length. These, as does the *common Cocoa-nut,* afford good examples of one numerous group of Palms which have their leaves *pinnated,* or divided like the plume of a feather. The two *stoutest* Palms in the collection, easily recognized by the thickness of their trunks and the great size of the tubs in which they are placed (each single plant, with its earth and tub, being calculated to weigh 17 tons), are the *West Indian* or *Jamaica Fan-Palms* (*Sabal umbraculifera*), a good example of a second extensive group, having *palmate* or *fan-shaped* leaves. The *Caryota urens* may be mentioned as deviating considerably in its foliage from other Palms: each leaf is very much divided, and the ultimate divisions or leaflets resemble in shape the fin of a fish.

We may further mention in this collection the *Date-Palm* (*Phœnix dactylifera*), producing the dates of commerce and of Scripture, and which, together with the *Dwarf-Palm* (*Chamærops humilis*), are the most northern of all Palms (the majority being tropical), extending even into the South of Europe; the *Guinea Oil-Palm* (*Elæis Guineensis*), which produces the African palm-oil; the well-known *Cocoa-nut* (*Cocos nucifera,* fig. 7.), of which the various

COCOA NUT.

uses, as fruit, milk, oil, wine or toddy, wood, fibre, &c., are said to be as numerous as the days in the year; the *Cabbage-Palm* (*Oreodoxa oleracea*), which yields the so-called esculent substance from the crown of its stem; *Seaforthia elegans* and *Corypha Australis* from New Holland; *Livistonia Borbonica;* and *Plectocoma elongata* from Dr. Wallich, which, with its luxuriant foliage, and its singularly spiny stem (the spines being digitate, or united together like the fingers of the hand, or still more resembling the foot of the mole, and admirably formed for strength), can hardly fail to attract the attention of the passer-by. Its leaves, when full grown, are of vast length, and pinnated like the shaft of a feather, so long, indeed, that they seem, as does the very slender stem, to need support; and nature has provided them with the means; for the rachis, or main-stalk of the leaf, extends, at the end, into a lengthened slender tail, armed all along with strong deflexed hooks, by means of which, while running up among the stems, and catching hold of the branches of other trees, the foliage and stem are propped. A yet more wonderful provision of nature is observed in the young and yet unfolded leaves of this plant, during the period when they insinuate themselves upwards among the branches of the forests, for then these spines are upright, and lie flat against the stalk of the leaf; not becoming reflexed till they are needed as a means of support. Of Arecas are the well-known *Areca Catechu* and *Areca sapida;* — *Sago-Palm* (*Arenga saccharifera*),—*Phœnix sylvestris* is the *Wild Date* of India, which yields palm-wine and sugar; — the *Ivory-Palm* or *Vegetable Ivory* is the *Phytelephas macrocarpa*, an inhabitant of the Magdalena, New Grenada, of which the seeds constitute a substance so exactly like ivory, that they have become a considerable article of commerce, and are used for turning into a vast variety of trinkets and other articles resembling ivory; and the *Wax-Palm* (*Ceroxylon andicola*, fig. 8. p. 22.), of the Andes of New Grenada, discovered by Humboldt, of which the full-grown stem is covered with a waxy substance having the same properties as bees' wax: —

and lastly, we may observe that many kinds will be seen to have a coarse fibre separating from the base of the leaves, so strong indeed, that in the *Attalea funifera,* and other Palms, it forms an extensive article of commerce from Pará, Brazil, for the purpose of making brooms and brushes, as well as for the machines employed in sweeping the streets of London and other cities.

8

CEROXYLON ANDICOLA.

Some or other of the *Bananas* or *Plantains* may always be seen in this House, in a more or less advanced state of flower or fruit, through the whole year, their ample and delicately green foliage overtopping many of the other plants. The clusters of blossoms form a long pendent spike, and the flowers are of two kinds; those which are situated at the base of the spike being destined to become the cucumber-like fruit, while the others, seen at the extremity of the spike, are covered with concave purple scales, that gradually drop off, and permit the escape of the pollen, or fertilizing dust, which, being conveyed by the wind or by insects to the other blossoms, renders them perfect. The *Banana* (fig. 9.) only differs from the *Plantain* (fig. 10.) in the form of the fruit: they are, indeed, considered by Humboldt as mere varieties. Both are of inestimable value to the inhabitants of tropical countries in the Old and New Worlds. A single cluster of fruit often weighs 70 or 80 pounds, even when produced in the stoves of

9

BANANA.

10

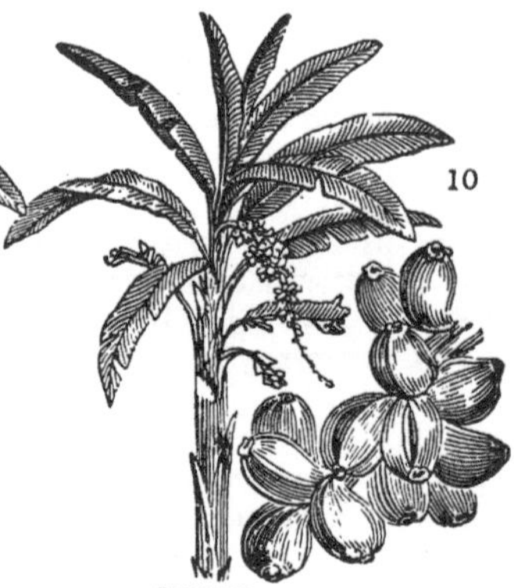

PLANTAIN.

this country. Besides being eaten fresh in their native land, bananas are dried as figs, or reduced to a kind of flour or meal by rasping. One kind, the *Plantain*, is called *Musa paradisiaca;* the *Banana* is *Musa sapientum.* A third and dwarf kind is the Chinese sort, *Musa Chinensis*, often called *M. Cavendishii.* The tender and succulent stems are eaten by various domestic animals: the fibre makes excellent cordage and clothing; and the leaves serve for covering houses. The rare and little known *Ensete* (*Musa Ensete*) of Bruce is another Musa, and for the knowledge and possession of it we are indebted to Walter Plowden, Esq., H. B. M. Consul at Mussowah in Abyssinia.

11

DRAGON'S-BLOOD TREE.

Two tall naked-stemmed plants in this House, with a crown of sword-shaped leaves, are the *Gum-Dragon Tree* (*Dracæna Draco*), which yields an astringent resin called dragon's blood, formerly used in medicine, and now chiefly employed by painters as a red varnish. Large as these specimens are, they are pygmies compared with the stature the tree attains in its native island, Teneriffe. "The gigantic tree of Orotava," (fig. 11.) says the enlightened traveller, Humboldt, "measures 45 ft. in circumference, a little above the ground." Tradition relates that this particular *Dracæna* was venerated by the Guanchos (the aborigines of Teneriffe), as was the *Elm of Ephesus* by the Greeks, and that in A. D. 1400 it was as large and hollow as it is now! Its growth being extremely slow, we may be sure the *Orotava Tree* is of incalculable age: doubtless it and the *Baobab* are among the oldest vegetable inhabitants of our planet. A noble specimen of a small branch of this tree is placed in the gallery of the Museum.—The classical Papyrus (*P. antiquorum*, fig. 12.). This is easily recognizable here by its tall, reed-like, triangular stem. It is crowned with the copious clustered flower-stalks. The stoutest individuals were selected by the ancients, and from the

12

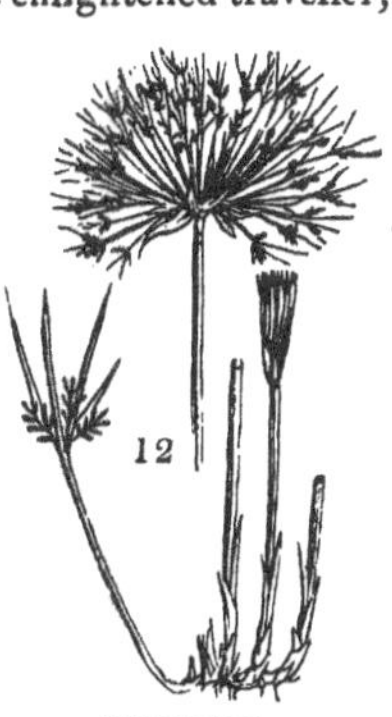

THE PAPYRUS.

white pith which fills the interior their paper was prepared. On this it is said that most of the old manuscripts are inscribed, especially those which have been brought to light by the excavations at Herculaneum and Pompeii. Specimens of ancient and modern paper made from this material are in the Museum.

The *Sugar-Cane* (fig. 13.), which happily can be no longer denounced, with regard to this country, as

> "The cane whose luscious juice supplies
> Europe's *blood-purchased* luxuries,"

distinguishes itself by its very large yet grassy character, long and pale-green foliage, and closely jointed stout stem. This latter, contrary to the character of most grasses, is solid (not hollow), and contains the saccharine juice, which is extracted by pressure between heavy rolling cylinders. The waste stems, thus squeezed dry, are generally used for fuel to boil the juice, and are found to be so impregnated with a siliceous or flinty substance, that masses of glassy slag are, in the course of a short time, deposited in the furnaces and require to be removed.

13

THE SUGAR-CANE.

The *Bamboo* (fig. 14.), when fully grown, is infinitely more gigantic than its ally the Sugar-Cane, attaining during one season, *in its native wilds*, a height exceeding 100 feet: its immense hollow stalks are applied to an infinity of domestic and useful purposes, as may be seen in the Museum.

14

THE BAMBOO.

The *Zamias*, *Cycases*, and *Encephalartus*, in the south wing of this House, are worthy of attention. Four of the finest were presented by Mr. Anderne and Mr. Moxon from the interior of South Africa. They are inhabitants of hot countries, chiefly in the southern hemisphere; and assuredly, within our days at least, nothing like them has ever been seen growing in temperate climates; but similar plants are found fossilized in the oolite formation of England, as at Portland Island, showing that in former ages these strange forms were denizens of this country! Their pinnated leaves are peculiarly harsh and rigid. The *Cycases* (fig. 15.) yield a kind of sago in the East Indies. Here are also noble specimens of the curious South-African *Elephant's Foot* (*Testudinaria Elephantipes*); so named from a resemblance in external

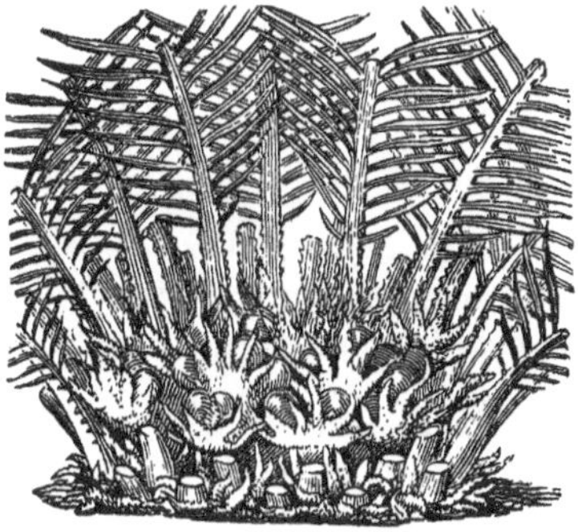

15 CYCAS REVOLUTA.

surface of the gigantic root-stock to the back of a tortoise, or to the foot of an elephant: and by the side of it is a plant greatly resembling it in appearance, from Mexico, but belonging to quite another sort of plant, the *Dioscorea macrostachya*, a kind of yam.

The *Calathea* (formerly called *Maranta*) *zebrina*, or *Zebra-Plant*, is easily known by its large beautifully striped copious foliage: each leaf is banded with shades of velvety green of different hues, and lined, as it were, beneath with purple. It is only by putting some of the leaves a little on one side that the clusters of purple flowers can be perceived. Here are seen two kinds of *Strelitzia:* one is *Strelitzia Regina* (*fig.*16.) 4-5 ft. high, showing its truly royal blossoms in winter and early spring; the other is the stately *S. augusta*, which, together with the Traveller's-tree, or *Urania speciosa* of Madagascar, have the most ample leaf-blades of any known plants.

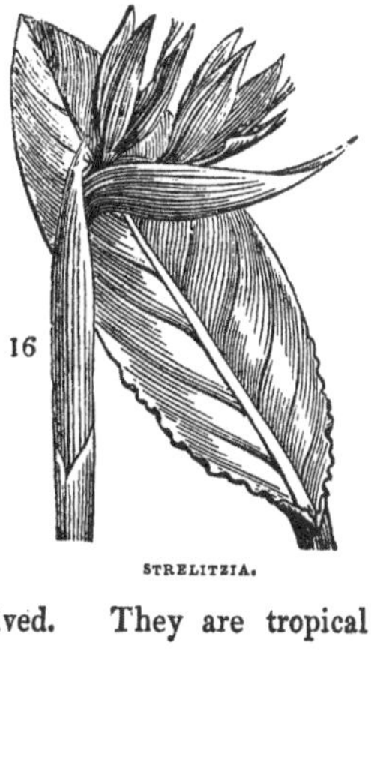

16 STRELITZIA.

Some very fine specimens of the *Screw-Pine* (*Pandanus*), are here rendered conspicuous by their tall, simple, or branching stem, and tufts of large leaves, somewhat resembling those of the Pine (i. e. *Pine-Apple*), but these spirally arranged; from which two circumstances the English name is derived. They are tropical

Indian plants, and are generally seen growing in Mangrove-swamps, by the sides of rivers influenced by the tides, and whence they would assuredly be washed away, but nature has provided that the stems and branches even should send down stout roots, which act as buttresses, and moor the plants, if I may so say, to their proper locality.

PAPAW.

Good plants of the *Papaw* (fig. 17.) and others of the *Chocolate Tree* (fig. 18.) are placed in this Tropical House. The juice of the former is employed in the East and West Indies for rendering tough meat tender; and, having this property, it is, of course, much prized by good housewives in climates where it is necessary to cook all animal food on the day when it is killed. A noble specimen of this, with a branching stem, and generally bearing fruit, has been recently presented to us by His Grace the Duke of Northumberland. From the seeds of the *Chocolate* Tree, as may be inferred from its name, is produced that "drink of the Gods," and also *Cocoa* (a very different thing from the fruit of the Cocoa-nut Palm, and a corruption of the Indian name *Cacao*, whence the botanical name, *Theobroma Cacao*).

CHOCOLATE TREE.

Of the genus *Euphorbia*, *E. grandidens* is here seen, with its lofty stout trunk, twelve or fourteen feet high, and sending out spreading whorled branches like a candelabrum. The slightest incision in the bark causes a great quantity of milky juice to flow, which, being of a highly acrid and venomous nature, is employed by the native Africans for poisoning their arrows and assagays. The juice of other allied species is used in various countries for intoxicating fish: a destructive mode of procuring the finny tribe practised in Ireland by poachers in the Shannon. The efficacy of our common *E. helioscopia* (*Wartwort*) in removing warts is well known in England.

Among other valuable trees in this House may be noticed the

Mango Tree (*Mangifera Indica*, fig. 19.) now annually yielding flowers, and sometimes its rare and delicious fruit; the *Silk-Cotton*

19

MANGO TREE.

Tree (*Bombax pentandra*); the *Longan* (*Nephelium Longan*); the *Mahogany tree* (*Swietenia Mahogani*, fig. 20.), a native of Honduras and Jamaica; the rapid-growing and thorny-stemmed *Ceiba* (*Bombax*

20

MAHOGANY TREE.

21

COFFEE TREE.

Ceiba). Here are the *Coffee-Tree* (*Coffea Arabica*, fig. 21.), seen in one place growing out of the crevices of the bare tufa rock of Bermuda, as obligingly sent by Governor Reid; the *Tanghin* (fig. 22.) or *Poison-Tree* of Madagascar (*Tanghinia veneniflua*), rendered infinitely more fatal than the Upas by the execrable laws of the Malagassy kingdom; the *Manihot* (*Jatropha Manihot*, fig. 23.), a most virulent poison, but whose roots (their deadly juices being removed by pressure or dissipated by heat) are made into the well-known *Cassava-Bread* of the West Indies, and into as great a variety of wholesome

22

TANGHINIA VENENIFLUA.

food as can be obtained from wheat; the *Cinnamon* (*Laurus Cinnamomum,* fig. 24.), whose bark constitutes the valuable spice so

23

JATROPHA MANIHOT.

named; and the *Bastard Cinnamon* (*Laurus Cassia*), of which the bark is said to be often substituted for that of true Cinnamon. Among the numerous kinds of *Figs* there will be found here, near the north entrance, a young plant of the *Banyan* (*Ficus Indica,* fig. 25.), one of the most celebrated trees in tropical India, for the immense stretch of its limbs and the singular mode provided by nature for their support: —

" Spreading so broad and long, that in the ground
The bended twigs take root, and daughters grow
About their mother-tree, a pillar'd shade
High overarch'd, and echoing walks between."—

24

CINNAMON.

These roots or props occupy such a space of ground in their native soil that one, growing on the banks of the Nerbuddah, covers an almost

BANYAN TREE. 25

incredible area, of which the circumference now remaining (for much has been swept away by the floods of that river) is nearly 2000 feet. The overhanging branches, which have not yet thrown down their props or supports, stretch over a much larger space : 320 main trunks may be counted, while the smaller ones exceed 3000, and each of them is continually sending forth branches and pendent roots to form other trunks and become the parents of future progeny. The whole (according to Forbes's "Oriental Memoirs," from which I quote) has been known to shelter 7000 men beneath its wide-spread shade. Our young plant, though it has already sent down many stout roots or props from its spreading branches, can of course give little idea of this famous tree ;

indeed, it is evident that a well-grown one would alone fill the entire Palm-Stove of the Garden. — The *Pepul Tree* (*Ficus religiosa*), from the same country, is remarkable for the tail-like points at the leaves; and these leaves abound so much in closely reticulated tenacious fibre, that the Chinese, by macerating them and removing the pulpy or parenchymatous substance, produce a kind of paper, which, when varnished, is capable of receiving the most beautiful drawings of birds, beasts, insects, flowers, &c. Such leaves, with the drawings, are commonly brought to this country from China, and are easily known to belong to this tree by their heart-shaped outline and the long tail-like point. Another kind of *Fig* in this collection must be here alluded to; for it illustrates a plant of Scripture, the *Sycamine-Tree,* or *Sycomore of Palestine* (fig. 26.), the tree into which Zaccheus climbed (*Ficus Sycomorus*): this is the true and original *Sycomore,* its name being derived from σῦκον, a *fig,* and μόρον, a *mulberry;* meaning a fig, whose leaves resemble those of the mulberry. " I was no prophet, neither a prophet's son," says Amos; " but I was an herdsman and a gatherer of *Sycomore fruit;*" from which, and from other passages in Scripture, it may be inferred that this tree was of very great importance among the Jews, although its fruit is extremely inferior to that of the true Fig (*Ficus Carica*), which two are the only eatable ones of 200 known species. The wood is said to be indestructible, and is therefore used for Egyptian mummy-cases. A fourth species of Fig-tree, the *Ficus elastica,* with large dark green glossy foliage, affords in its milky juice the *Caoutchouc* of the East Indies.

26

SYCOMORE.

But we must proceed; and, as space will not permit the mention of a tithe of the interesting plants in this stove, we content ourselves with saying that here may be inspected, flowering at some period or other of the year, a great variety of tropical shrubs and trees; and amongst them the feathery foliage of the *Tamarind-Tree* (*Tamarindus officinalis*), whose preserved fruit is an extensive article of commerce; the *Cotton* (*Gossypium herbaceum,* fig. 27.), the seeds of which are surrounded by that beautiful filamentous substance, and whose flowers resemble those of a *Hibiscus*; *Indigo* (*Indigofera Indica,*

27

COTTON TREE.

fig. 28.), the leaves of which yield the rich dye so called; and the great strangely-shaped lurid and fetid flowers of the *Aristolochia ornithocephalus*. Lastly, we shall only mention in this stove the numerous climbers planted in the ground at the base of the pillars and of the stair-case, such as *Convolvuluses, Passion-flowers, Aristolochias, Bauhinias, Telfairia, Poivrea*, &c., all remarkable for the beauty of their foliage or flowers, and sometimes of both.

28

INDIGO.

Impossible as it would be to record, in this brief Guide-book, the numerous donors of rare exotics to this Establishment, it is only our duty to state that, of the kinds from the East Indies, in the Palm as well as in other tropical Houses, by far the greater number were sent from the Honourable the E. I. Company's Botanic Garden at Calcutta, by the late distinguished superintendent, our lamented friend Dr. Wallich. The Books of that garden, as proved by a Report now before us, printed at Calcutta in 1840, show that, in the five previous years alone, 9 Cases, with 229 plants* of the rarest and most valuable description, were transmitted here; and between 1840 and 1845, the period of that gentleman's retirement from his arduous duties, our own Books testify to the arrival of 13 Cases containing 275 plants!

Quitting the Palm-house, we find that, from the south-east angle, the walk extends round the water; and from the opposite side of that piece of water (near the NEW MUSEUM †) the best view of the structure may be obtained, and in calm weather its reflection is seen in the lake. Continuing past a wooded hill, the visitor is recommended to direct his steps north, and he will soon catch sight of a portion of ground recently laid out for the HARDY HERBACEOUS COLLECTION, and containing a small building also fitted up as a MUSEUM, and several stoves and green-houses, hereafter to be briefly noticed. This ground can be visited in this direction and the MUSEUM inspected, and the principal cluster of Plant-houses (Nos. 10, 11, &c.) thus approached; or, if the visitor prefers another route, he may leave the herbaceous ground and the end of a long wall on the right, near which is a fine *Cryptomeria japonica* (first raised from seeds

* The same Report further states, that during the same period there were 2107 applicants for plants to the Calcutta Garden, from different parts of the world, who were supplied to the enormous extent of 189,932 individual plants. It is to be regretted that this document, printed at Calcutta, has not been more generally circulated: for it affords valuable information, relative to the introduction as well as distribution of a great number of rare and useful plants, during a small portion only of that gentleman's able directorship.

† See "Museum-Guide," 3rd edit., for a description of the contents of this and of the *Old Museum*.

sent from China by the late Sir Everard Home), passing a stately *Tulip-Tree* on the left; and, taking the next turn to the right, he will come to a plot of ground now a medical garden. In this quarter, too, he will find two good but young trees of the Paper Mulberry (*Broussonetia papyrifera*), whose bark yields the tapa or clothing of the inhabitants of the South Seas; and the most beautiful and almost gigantic Pampas grass (*Gynerium argenteum*), 11 feet high when flowering in the autumn; its large panicles of flowers very much resemble those of the sugar-cane. Though a native of the Pampas of Buenos Ayres, it is quite hardy. Just beyond is a group of four low Plant-houses; the first is

No. III. MESEMBRYANTHEMUM HOUSE,

a *small* GREENHOUSE, during the winter occupied with a considerable collection of African plants, of the genus *Mesembryanthemum* (or *Fig-Marigold*), but placed in the open air in summer. Many of these are remarkable for the resemblance in their foliage to the jaws of animals, whence some are appropriately named *felinum, tigrinum, caninum, vulpinum,* &c. The capsules of others have the same hygrometric property as the entire plant of the famous *Rose of Jericho*, or the hygrometric *Club-moss* (of all of which examples may be seen in the Museum); for, contrary to the nature of capsules or dry seed-vessels in general, these open in wet weather into segments, resembling the petals of a flower, and close in dry,—a beautiful provision of nature, by which the seeds sow themselves at the only season suited, in those hot sandy deserts, to their germination;—and, after being gathered, they long retain this property, and may be made to open or shut according as they are placed in a wet or dry atmosphere. In the summer these *Fig-Marigolds* are removed to the open air, and the House is then occupied with Fuchsias or other showy and ornamental plants.

No. IV. DOUBLE ORCHIDEOUS HOUSE,

is a *low double* STOVE, chiefly occupied as by Orchideous plants recovering from the effects of a long voyage, or used as a nursery for bringing forward flowering specimens for the more ornamental houses.

No. V. ORCHIDEOUS HOUSE,

a long *low* Stove, now wholly occupied as an Orchideous-house; but as the House No. 4. is also devoted to the same family of plants, and as there is often a necessity, from increased growth or other circumstances, to remove certain kinds from the one house to the other, our notes upon them must include the united collections.

The collection is eminently valuable. Of late years, there have been added to the original collection the extensive one of the late Duke of Bedford, presented by Her Majesty Queen Victoria, and the equally valuable legacy of the late Rev. John Clowes, M.A., of Broughton Hall, Manchester, in the autumn of 1846; together with many species, procured by purchase, or from our collectors and friends abroad, and other sources. The tropical *Orchideous Plants* (or *Epiphytes*, as most of them are justly termed, from the fact of their being generally found growing on the trunks and branches of trees) are at this time the greatest favourites among cultivators; and the prices given for many would surprise any person not individually interested in them. When in flower they are certainly among the most beautiful objects of the vegetable creation, and remarkable for their highly varied and peculiar forms, great delicacy of texture, and often exceedingly brilliant colouring. Happily, too. there is not a month nor a day in the year that some or other is not in blossom; though the first powerful suns of spring induce the flowering in a very marked degree. Many are here seen attached to branches of wood, or placed in wire baskets with moss and bark, or planted in the husk of cocoa-nuts and suspended from the rafters, living, as it were, and flourishing on heat and moisture. It were an endless task to direct attention to any particular kinds; for their beauty depends on the presence or absence of blossoms, which, with some exceptions, are generally short-lived. Still we may observe in the collection the "*El Spirito Santo*," (*Peristeria*), *Cattleyas*, *Dendrobia*, *Stanhopeas* (exhaling, as do many others, a most powerful fragrance); *Lælia superbiens*, remarkable for the large size of its flowers; *Phalænopsis amabilis*, or *East Indian Butterfly-Plant*, with its corollas of the purest white; and *Oncidium Papilio*, or *West Indian Butterfly-Plant*, whose resemblance to an insect is increased by the presence of certain petals, which look like long antennæ, and by the flower being borne on a long slender stalk, far away from the leaves, and which seems to carry the fly into the air. The *Vanilla* (*Vanilla aromatica*, fig. 29.) is one of these tropical *Orchideæ;* its long narrow pods (not unlike those of the Haricot bean, but dark brown in colour, and soft and oily to the touch) afford the fragrant *vanilla* of commerce, much used in the preparation of chocolate, and in various other ways as a condiment, being considered to promote digestion. It is a native of the hot parts of South America, and from Vera Cruz alone the pods of this plant are exported to the amount of 40,000 dollars annually. Our largest specimen of this, however, will be found in House No. 15.; it was presented in 1854, with many other rare plants, by His Grace the Duke of Northumberland. Our species of the East Indian genera *Aërides* and *Saccolabium* are among the most highly prized. At the latter end of winter, the very lovely East Indian kinds of *Cypripedium* are generally in blossom; these, however, with some others, are terrestrial, and planted in pots of soil. Of this terrestrial kind is the *King-Plant* of the Cingalese (*Anœctochilus setaceus*), as rare as its leaves are beautiful: the foliage closely resembles brown-green velvet, with the most exquisite net-work of gold; other kinds (*Physurus* of Brazil) have green leaves reticulated and spotted with white.

29

VANILLA AROMATICA.

No. VI. VICTORIA HOUSE.

Here, in the summer and autumn, will be seen the *Victoria regia*, described at p. 17, and various other water-plants. Two aquatic *floating* plants are worthy of inspection. One resembles bright-green lettuces floating on the surface: it is the *Pistia Stratiotes* of the West Indies, the *Duckweed*, in fact, of tropical countries, resembling green lettuces. The other is distinguished by the leaf-stalks being remarkably inflated, with large air-cells within; these give buoyancy to the plant, which would otherwise sink: it is the *Pontederia crassipes*, and bears a beautiful blue flower. The shelves contain a very choice assortment of plants of hot countries suited to the atmosphere of this House, and many remarkable for the beauty of the flowers: *Ixoras*, scarlet *Clerodendrons*,

Medinilla magnifica, and *M. speciosa*, *Torenia asiatica*, *Hoya imperialis*; *Aristolochia grandiflora*, with the huge blossoms of which, Humboldt says, the children adorn the head, in lieu of a hat or bonnet; *Passion flowers*, among them the *Grenadilla*, *Æschynanthuses*, &c.

Here, too, may be seen a group of different plants, especially characterised by the varied colouring or marking of the foliage, often called "painted-plants;" among them the *Caricature Plant* (*Graptophyllum hortense*), many of the spots of whose leaves bear a very accurate resemblance to the human face, more or less divine. The fragrant *Lemon-grass* will be seen here (*Andropogon Schœnanthus*), admirably figured and described in Dr. Wallich's superb "Plantæ Asiaticæ Rariores," where the author observes, "This is a favourite herb with the Asiatics both for medicinal and culinary purposes, and is found to afford a drink generally very grateful to the palate in sickness. Dr. Maton, Physician Extraordinary to the late Queen (Charlotte), informed me that he had been repeatedly treated with a dish of Lemon-grass tea by her Majesty, who used to be very fond of it, and was supplied with the plant from the Royal Gardens at Kew." Its fragrance is exactly that of Lemon or *Verbena triphylla*. Here, too, the *Patchouli* or *Pucha-pat* of India (*Pogostemon Patchouli*), the most esteemed perfume of the present day. In a large earthen pan with water, may be seen here one of the most wonderful of all vegetable productions, the *Lace-* or *Lattice-leaf* of Madagascar (*Ouviranda fenestralis*) recently brought to us, living, by the Rev. W. Ellis. Its leaf is formed of longitudinal and transverse fibres, having no parenchyme or cellular substance in the interstices, which are thus open like the lattice of a window, or a piece of lace.

30

AUSTRALIAN PITCHER PLANT.

At the time we write may be found in this stove some rare Malayan Pitcher plants, though not so large as those in stove No. 11. (p. 45.); the beautiful and much rarer *Australian Pitcher-Plant* (*Cephalotus follicularis* of Brown, fig. 30.); and the still more singular *Dionæa muscipula*, or *American fly-trap* (fig. 31.), which has at the end of each leaf, as its name implies, a veritable living trap, consisting of two broad fleshy lobes, jointed in the middle, fringed with long spines, and furnished with two or three hairs on the disk of the lobes. The moment an insect (or any extraneous body) touches the hairs on the disk, the two lobes close firmly, and press the luckless intruder to death: the struggles of the victim only occasioning the lobes to shut more firmly, by which its destruction is hastened.

As soon as the insect ceases to struggle, and dies, the trap opens, ready to continue the work of destruction; but there is no reason whatever to suppose that the dead insects in any way nourish the plant. Equally remarkable in the structure of the foliage are the *Sarracenias* (fig. 32.), or *Side-saddle flowers*, so called from the form of the stigma: they have tubular leaves containing a fluid, and inverted hairs at the mouth of the tube. Insects, in the native country of these plants especially, and not a few with us, are attracted by the fluid: the inverted hairs hasten the descent of the intruder, which falls into the fluid without chance of escape, and perishes.

31

AMERICAN FLY-TRAP.

We may now proceed westward; and observe a fine Hop-hornbeam (*Ostrya vulgaris*); further west is a noble tree of the *Sophora Japonica*. We may turn to the right, towards the principal cluster of Plant-houses, and we observe on the left a large harsh, rigid-looking *Pine*, which cannot fail to attract the attention of the visitor: it is the famous *Chili Pine* (*Araucaria imbricata*), brought to England in the year 1792 by Mr. Menzies, the surgeon in Captain Vancouver's voyage: it now frequently bears its curious cones, but they never ripen with us. It is, perhaps, not generally known, that the seeds are eaten for dessert in Chili, as are those of the Stone Pine (*Pinus Pinea*) in Italy, and almonds with us. Westward of the *Araucaria*, upon the same piece of lawn, and forming a striking contrast by its gracefulness, stands a splendid tree of the *Weeping Birch* of Scotland. From this spot we shall find it convenient to enter

32

SARRACENIA.

No. VII. SUCCULENT HOUSE.

This is an excellent house of its kind, 200 feet long and 30 feet wide, recently (in the spring of 1856, completed, and destined for those plants, chiefly from warm (but not quite tropical) and dry countries, which are denominated by horticulturists *Succulent Plants* (*Plantes grasses* of the French). They include *Cactuses*, *Aloes*, *American Aloes*, many *Euphorbias* (*Agaves*), and other Amaryllidaceous plants, *Crassulas*, *Sempervivums*, *Stapelias*, &c. &c. Many of them are eminently worthy of notice.

1. The collection of *Cactuses* is perhaps unique of its kind, thanks to various friends in the warmer parts of the New World, of which countries they are exclusively natives. If these plants do not possess much grace and beauty (their flowers, however, are often splendid in the extreme), yet they are very remarkable for the strange forms and structure of their, almost invariably, leafless stems and trunks, their deep longitudinal ribs or furrows and sharp angles, the singular vestiture of hairs or spines (or both combined),—the latter, often in countless myriads, are arranged with the most perfect symmetry in stellated or star-shaped clusters, sometimes not thicker than bristles, and scarcely two lines long, or they are broad and transversely banded like lobsters' horns, at other times long and straight, and so strong as to serve the Mexicans to fasten their "ponchos" about their persons. Some species resemble the convolutions of the brain. One set is distinguished by their tall and curiously jointed and flattened stems; they are the *Opuntias* or *Nopals*, some of which yield the fruit much eaten in warm countries, under the name of *Indian Fig* or *Prickly Pear* (*Cactus Tuna*, fig. 33.), and their stems are used for making almost impenetrable fences (these are

33

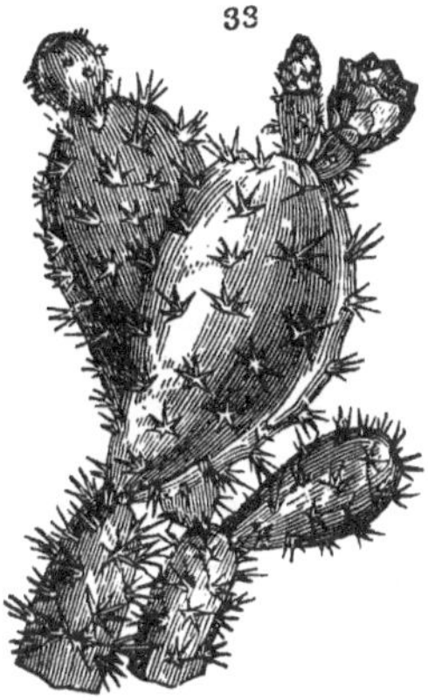

CACTUS TUNA.

34

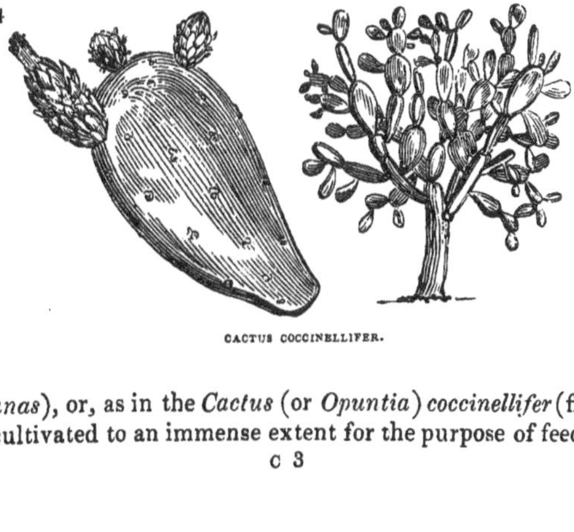

CACTUS COCCINELLIFER.

called *Tunas*), or, as in the *Cactus* (or *Opuntia*) *coccinellifer* (fig. 34.), they are cultivated to an immense extent for the purpose of feeding the *cochineal insect,* a small kind of meal-bug (seen in this house), reared in such quantities, that, from Mexico alone, Humboldt assures us 32,000 arrobas of cochineal are annually exported, equal to 500,000*l.* sterling. The insect has the power of extracting the juices, and converting them, by a chemical process, into the rich scarlet dye called *Cochineal;* but it is not so generally known that the *fruits* of the *Nopals* secrete the same colour, and excellent cochineal has been obtained from the fruit, as well as from the insect, in the East Indies. Another tribe of *Cactuses* resembles in form and spinous character the Sea-Urchins of our shores (*Echinus*), and is appropriately denominated *Echinocactus.* Here are good specimens of the *Echinocactus Visnaga* so called (*Visnaga* meaning a toothpick in Spanish) from the use made of the spines in Mexico; for which we are indebted to Frederick Staines, Esq.*

Here is a lofty specimen (together with many lesser ones) of the *Old Man's Head Cactus* (or *Cereus senilis*), 14 feet high. This species is called *senilis,* from the quantity of long wiry grey hair which crowns the summit (particularly observable in the younger plants). Unlike the human kind, the old plants are less conspicuous by their grey hair than the young ones. For the means of procuring them we are indebted to John Taylor, Esq., and the Directors of the Real del Monte Company. *Small* plants of this species we know to be twenty and twenty-five years old: from their slowness of growth, as well as from the reports of the inhabitants of Mexico, there is reason to believe that these gigantic individuals are some hundreds (probably a thousand) of years old.

35

GROUP OF ALOES.

2. The *true* or *African Aloes* (fig. 35) may next claim atten-

* Here we are anxious to record that the same gentleman did us the favour, with infinite labour, to send us a much more magnificent plant of this Cactus in 1846, and weighing one ton! It was planted, and looked sound and perfect, and was, for some months, the wonder of the Garden; and it found a place in the "Illustrated London News" of the day, where an excellent representation is given; but it had met with injury during its perilous journey or voyage; a bruise appeared; and decay soon extended through the whole of this enormous mass, tainting the air with its fetid smell

tion, of which more than one kind yields the medicinal *Aloes* of the shops. They are peculiar to Africa, and form a striking feature in the vegetation of South-eastern Africa, one species the *Socotrine Aloe,* extending to the island of Socotra.

Of the same natural family as this true Aloe is the *New Zealand Flax* (*Phormium tenax*, fig. 36.) Its leaves are like those of our *Iris*, or flag, and abound in a strong fibre, which recommends it for an immense variety of purposes where hemp or flax would be used in Europe.

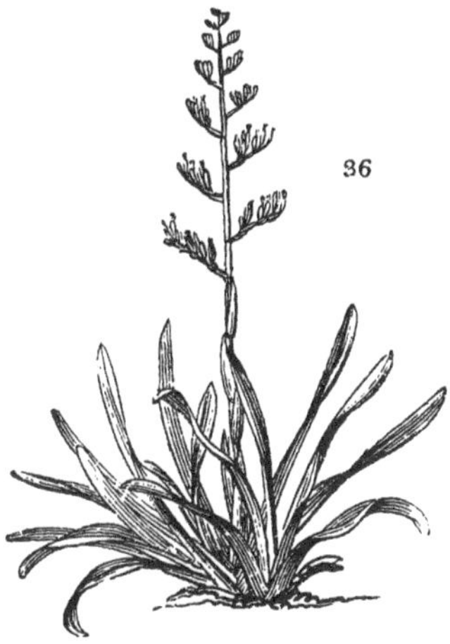

36

NEW ZEALAND FLAX.

3. Another extensive group of succulents is the *Agave* or *American Aloe Family*. The *Agave Americana* is familiar to most frequenters of gardens. It blossoms, not, as the story goes, only once in a hundred years, but, in reality, more frequently than other kinds, and throws up a flower-stem 20 to 24 feet high. The Plant-house we are now surveying contains one very large unknown species of *Agave* (marked "*Agave, Mexico*") with huge, sword-shaped, thick, and fleshy leaves: each of these leaves averages 12 pounds in weight. Two ancient specimens of another and not smaller sort, but which, we regret to say, can now only be seen in a young state, blossomed here in 1844, and attracted much attention: we mean the *Agave vivipara* of Linnæus (*Fourcroya gigantea* of modern authors, a name given in compliment to the French chemist, *Fourcroy*). The two plants in question had been in the Royal Gardens, first of Hampton Court and then of Kew, probably from the earliest introduction of the species into Europe, upwards of a century ago (in 1731). On one and the same day, in the summer of 1844, each was seen to produce a flowering-stem, which resembled a gigantic head of asparagus, and grew at first at the astonishing rate of two feet in the twenty-four hours. So precisely did the twin plants keep pace with each other, that at the very time it was found necessary to make an aperture in the glass roof of the house for the emission of one panicle of flowers (26 feet from the ground), a similar release was needed by the other. The rate of growth then most sensibly diminished; still, in two months, the flower-stalks had attained a height of 36 feet!

The flowers were innumerable on the great panicles: they produced no seed, but were succeeded by thousands of young plants, springing from the topmost branches (whence the Linnæan name of *vivipara*); and these continued growing while attached to the stem for a long while after the death of the parent-plants, both of which perished, apparently from exhaustion. Our collection now, therefore, contains young individuals chiefly of this particular *Agave.* All of this family yield fibre for cordage, cloth, &c., and the pulp is fermented and becomes an intoxicating drink.

4. A small collection of *Stapelias* or *Carrion-flowers* (fig. 37.) of South Africa, is here seen, whose quadrangular leafless stems bear indeed some similarity to Cactuses, while their flowers resemble star-fish. Their odour is such that flies, attracted by it, lay eggs upon them in great numbers, taking them for putrid meat: the poor larvæ, when hatched, find the difference, for there is nothing for them to feed upon, and they perish in great numbers; thus it would appear that *Stapelias* are among the many plants destined by the Author of nature to keep insect-life within due bounds.

37

CARRION FLOWERS.

5. Here is a remarkable plant from South Africa, a kind of *Geranium* (or *Monsonia*), remarkable for its resinous quality *when dead*, the *Monsonia Burmanni* (*Geranium spinosum* of Linnæus), of which the quantity is so great in the burning sands that it has been imported, in the hope of its proving valuable as an article of commerce. This particular plant, on its arrival, and for four years, had been to all appearance perfectly dead, and more than half converted into a gum-resinous substance, exhibiting only a few crooked dead-like branches. Suddenly, in the spring of 1850, it put forth leaves, and is full of life and vigour. (Only one plant has since died, and is preserved in the Museum.)

6. This house contains some leafless (and other) *Euphorbias* (figs. 38, 39.), whose forms a good deal resemble those of certain Cactuses, but whose milky juice is eminently poisonous, and extensively employed in South Africa for rendering mortal the wounds of arrows and assagays. *E. splendens* and *E. Bojeri* are distinguished by their vivid scarlet flowers.

7. Some *Sempervivums* of the Canary isles are remarkable for their flattened crown of leaves; and lastly, we can only mention that

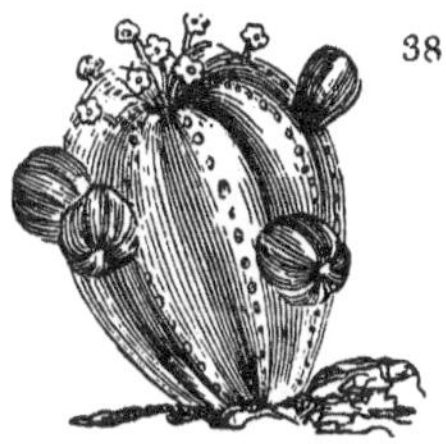

38

EUPHORBIA MELOFORMIS.

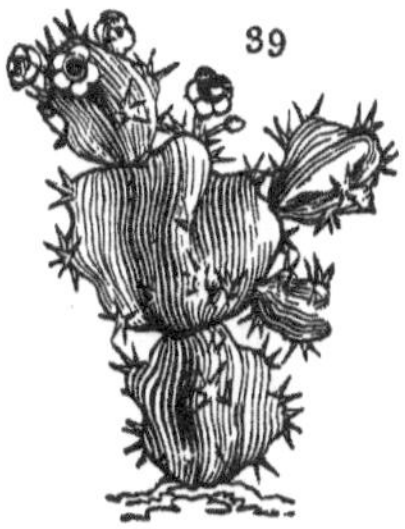

39

EUPHORBIA ANTIQUORUM.

there are some large *Bromeliaceous* and several *Amaryllidaceous* plants allied to the *Agaves* and others, though not very succulent in themselves.

Leaving this House by the western door, and looking north, we see the

HARDY AQUARIUM

(marked O on the plan), a stone tank of bog and water, containing hardy aquatic plants of England and other cool countries: among the more interesting are a few from the Falkland Islands and Tierra del Fuego, particularly the celebrated *Tussack-Grass* of the Falklands (*Dactylis cæspitosa*), introduced by Sir James Clark Ross and the officers of the Antarctic voyage. It is one of the most valuable coarse agricultural grasses, and, having braved the droughts and cold and heat of England for several years, there is no question that it may, with care and patience, be naturalized. It is slow of growth, and slower to form its great tussacks, whence is derived the name. They, together with the mass of foliage, constitute thickets where wild cattle and more wild runaway sailors find shelter and protection, and both obtain *food;* for it is related, by the late governor of those islands, that two runaway sailors for a long time subsisted on the raw young shoots of this grass, which are, moreover, brought to table, boiled like asparagus. Indeed the *Tussack-Grass* is now flourishing luxuriantly in the Orkney Islands and Hebrides, especially in Lewes, under the auspices of Sir James Matheson, M.P.; vicinity to the sea, an equable climate, and cool atmosphere being essential to it. Close to this tank is an entrance, by the western door, to

No. VIII. THE NEW ZEALAND AND CONIFEROUS HOUSE.

This venerable structure, erected by Sir W. Chambers in 1760, has seen its best days, and, judging from the favourite reception lately given to the director's report in the House of Commons, we are warranted in believing that ere long instructions will be given by the Chief Commissioner of Works to replace it with another in the grounds more worthy of this establishment, and more suited to the purpose for which this is now destined, namely, the reception of *Trees and Shrubs of temperate climates*. The present contents in winter * are mainly *Conifers* and *New Zealand Plants*, but they are by no means confined to them; the collection is miscellaneous, and larger specimens of the same kinds, or some of them, are repeated in the ORANGERY (Plant House No. II.). Here are young plants of the *Cowdie* or *Kauri Pine* (*Dammara australis*), the *Dacrydium cupressinum*, whose feathery boughs perhaps exceed in delicacy and grace those of all other forest-trees; the *Celery-leaved Pine* (*Phyllocladus trichomanoides*); the very rare *Thuja Doniana*, Hook., a most elegant new *Arbor vitæ* of New Zealand; the *Mai* or *Metai* (*Podocarpus spicata*), and the *Miro* or *Mairi* (*P. ferruginea*); the *Mammoth tree* of California, *Wellingtonia gigantea;* together with the singular *Aralia crassifolia*, a kind of ivy, bearing long leaves, of a texture almost resembling whalebone. Here, too, are the *Myrtus bullata* of *New Zealand*, with its blistery leaf; the charming *Metrosideros robusta*, which climbs over other trees like ivy, and adorns their otherwise bare trunks with its large glossy foliage and brilliant scarlet flowers; and the *New Zealand Beech* (*Fagus fusca*).

Here grow the two rare *Beeches* of Tierra del Fuego, *Deciduous Beech* (*Fagus antarctica*), and the *Evergreen Beech* (*Fagus betuloides*), the latter remarkable for its beauty and small evergreen foliage, scarcely larger than that of the broad-leaved myrtle, and for its being the most southern-growing tree in the world; indeed, but little vegetation of any kind exists beyond it. Its size and form, however, in its native region, depend on the place of growth. In sheltered valleys it attains a considerable size, with a trunk 7 feet in diameter, so that Captain Philip King made large boats from one trunk; while on the exposed heights of Hermite Island the trees are so dwarfish and stunted, and the branches so densely compacted, that the traveller is able literally to walk upon the tops of them! (One has been planted in the open lawn, and has stood the winter well; but we fear the summer's drought for it more than the wet or cold of winter.) Near them stands another rare *Evergreen Beech*, the *Fagus Cunninghamii* of Van Diemen's Land, and the still rarer *Huon Pine* (*Dacrydium Franklinii*) of the same country. In the spring and summer a delicious pine-apple like fragrance is often perceived by the visitor at

* One end of this House is temporarily occupied with a noble *Tree-fern* of New Zealand (*Cyathea medullaris*), recently presented by H. R. H. the Prince Albert.

the west end of this House: it is diffused by the blossoms of the Chinese *Magnolia fuscata.* Young plants of the *Argan* tree (*Argania Sideroxylon*), of Morocco, is placed in this House. It possesses no beauty, but its fruits are very valuable, in the husks for feeding cattle, and in the nut for the copious oil. The wood, too, is beautiful.

Here will be found several South African plants, some *Proteaceæ* * of the Cape colony, in which plants the Garden was formerly very rich; but they gradually died out; and, strange to say, although the botany of South Africa has been of late years investigated beyond that of every other part of the world, seeds and living plants have been almost wholly neglected; so that in the plants of no country are these and other European gardens more deficient than in those of South Africa. An idea may be formed from the woodcut (fig. 40.) of the beauty of some that have formerly flourished in this Garden, few now remaining, and of all of which we should be thankful again to receive seeds. The most beautiful of the Cape *Proteaceæ,* namely, *Leucodendron argenteum,* is the common firewood of the colony. *Protea mellifera* and others yield a honey which is boiled down and used in coughs. In Chili, one species, *Quadria heterophylla,* yields an esculent nut, sold in the markets under the name of *Avellano:* and this will be found in the present House.

40

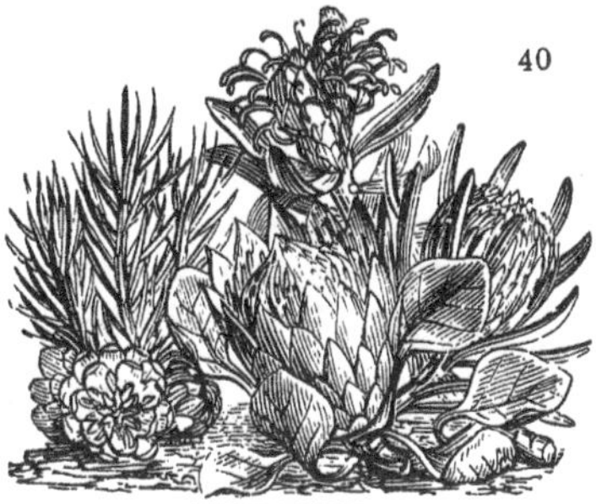

CAPE PROTEACEÆ.

Gnaphaliums and *Xeranthemums,* shrubby *Everlasting-Flowers* (fig. 41.), are so called from the nature of their blossoms, which neither shrivel, nor, for a long time after being gathered, lose their brilliant colours. Not unfrequently, in one part of this House, a powerful foxy odour is perceptible. This arises from several kinds of *Diosma* or *Bucku,* a favourite cosmetic of the Hottentots in S. Africa, who mix it with grease, and smear it over their persons to keep away flies.

41

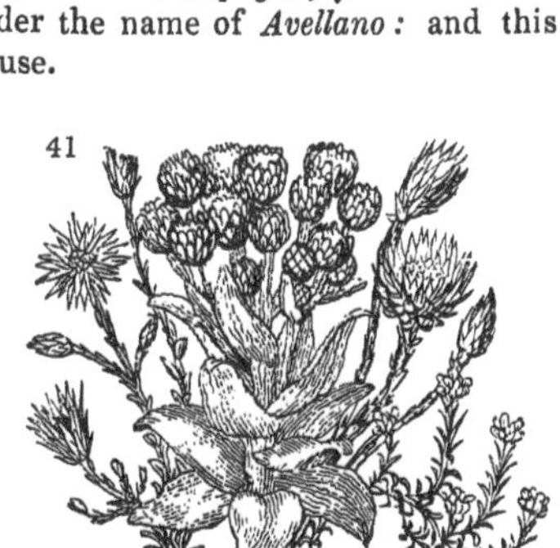

GNAPHALIUMS AND XERANTHEMUMS.

* Now removed to the Conservatory, No. 1.

Quitting this House at the eastern door, and passing an oval bed, filled with the exotic-looking yet hardy *Yuccas* (or *Adam's Needles*), the nearest Plant-house is a low building with a span roof, of which the entrance is at the east end; viz.

No. IX. A PROPAGATING HOUSE,

with a double span roof. Unless it contains, as at some seasons, any thing of peculiar interest, it is kept private, and is now almost entirely occupied with numerous young Palms from various countries, and Ferns, and seedlings of various sorts, aquatics, *Victoria regia*, *Nymphæas*, &c., and young plants of the rare *Doum Palm* (*Hyphæne Thebaica*, from Upper Egypt, sent home by Dr. Hooker), and the still rarer *Hyphæne coriacea*, from South-eastern Africa; but both kinds are most difficult to rear. Opposite to the door is an entrance to

No. X. THE AUSTRALIAN HOUSE,

a cruciform structure of large dimensions and excellent arrangement, is at all times occupied by plants of interest and beauty, and if the visitor happens to come at the latter end of winter, or in early spring, he cannot fail to be struck with the variety and fragrance of the inmates of this building. It is filled at those seasons with an unique collection of plants of Australia.

42

NEW HOLLAND ACACIAS.

The *Leguminosæ* are in perfection during the early spring months; then the fragrant *Acacias* (fig. 42.) are eminently worthy of a visit. The *New Holland* Acacias, as is well known to naturalists, exhibit a remarkable conformation of foliage. In other countries their leaves are perfect, having the normal character, more or less compound and pinnate, with numerous leaflets: but in the innumerable species from Australia (with some exceptions) the seed-leaves only are compound; as they advance in age they cast off the leaflets, and at length the plant produces only *leaf-stalks*, which widen, and have the appearance, and perform the functions of, true leaves. These leaf-stalks (called *phyllodia*) are easily recognised by their position: it is not the *flat* surface, but one of the *edges*, which is vertical, or directed to the zenith.

In the summer, these Australian plants are removed to the open air, the House is filled with various plants of a still more ornamental character, and of the greatest floral beauty.

No. XI. THE TROPICAL FERN HOUSE,

is situated due north of, and close to, the Australian House. It is hence most conveniently approached from the east. Passing through a small porch, a span-house is entered, 90 feet long and 28 feet wide, glazed with sheet-glass, having a double slate staging in the middle, facing north and south, with a walk through the centre, and another walk on the outside and around the staging, with stone shelving under the walls. This fine area is occupied with *Ferns* to which the heat and moisture of this place are suited. The *Ferns* constitute a very valuable collection, and nothing can exceed the variety, beauty, and elegance of the leaves or fronds. Some of the smaller and finely-divided kinds are clothed on the underside with a delicate yellow pulverulent substance, others with a white powder. The "*Great Stag's Horn Fern*" (*Platycerium grande*) is one of the most remarkable in appearance and form, together with *P. stemmaria* and the well known *Pl. alcicorne,* or *Common Stag's Horn Fern.* The first (*Pl. grande*), a native of Australia, where it grows on the trunks of trees. The *young* stage of it is a small green leaf or frond lying flat against the soil, lobed and divided at the margin something like a stag's horn. It thickens with every succeeding growth of leaf, and this addition is alternately right and left over the older leaves, which die and contribute to the nourishment of the plant. These are the sterile fronds. The plant bears another kind of frond, projecting and even drooping, narrowed at the base, broad and two-lobed above, and *beneath* bearing a great brown spot of fructifications. This rare plant was presented to us by the late Mr. Bidwill. A second plant he sent over was purchased by the late Duke of Northumberland for twenty-five guineas. Larger, and especially "*Tree*" *Ferns,* which cannot be accommodated in this House, will be found in the adjacent one, No. 12., which communicates with it.

In this tropical FERN-HOUSE, as admirably promoting their health and vigour, are our finest and oldest specimens of *Pitcher-Plants*, the *Nepenthes distillatoria* (fig. 43.), and the infinitely rarer *N. Rafflesiana:* the latter was successfully brought home from Singapore in a Ward's-case by Captain Bethune, R.N., and by him presented to the Royal Gardens. Both species are more or less scandent. The leaves are terminated with an appendage exactly resembling a pitcher, of considerable size, having a lid at the top. When young, the lid is firmly closed, yet, even at that period, it contains a considerable quantity of fluid, distilled, as it were, by the plant (whence is derived the name of the species in more general cultivation): after a time the lid opens, and continues firmly attached at the back of the orifice by a hinge, and never again closes. With us in the summer season, and in its native Malayan islands at all seasons, insects visit the pitchers in great numbers to get to the liquor, fall in, and from the difficulty of escape are drowned, sometimes filling the entire cavity. — Other lesser species of *Pitcher-plant* are seen in House No. 6.

43

PITCHER PLANT.

No. XII. AMHERSTIA HOUSE.

So called because it contains the noble specimen of *Amherstia nobilis* (presented by the late Mrs. Lawrence), of which younger plants are here also. It is occupied by a miscellaneous collection of tropical plants, and among them our finest *Ferns* occupy a conspicuous place, especially (though young) *Tree Ferns*, which are among the rarest and most valuable; of all plants they are the most difficult to import alive, except while very young. In *our* country, *Ferns* are of humble stature, their leaves, or fronds, emerging directly from the ground; but in the *tropics*, and even in the more temperate parts of the southern hemisphere, these fronds, 15 to 20 feet long, are elevated on unbranched trunks (resembling Palms), 20 to 40 and 50 feet high. Here are splendid specimens of the rare Fern *Angiopteris evecta*, with its gigantic foliage, and curiously knotted short trunk.

A portion of the side-shelves in one part of the House is occupied with an extensive collection of the genus *Begonia*, whose highly ornamental foliage preserves, "amid a hundred modifications," its

44

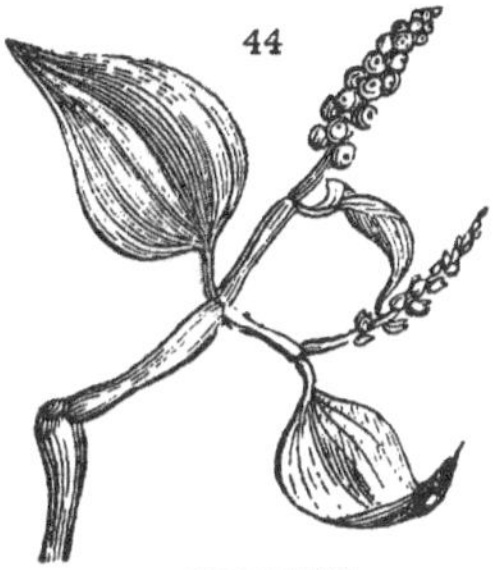

PEPPER PLANT.

peculiar character of obliquity, whence the plants are not inaptly named *Elephant's Ears*; the species, too, possess a great recommendation in producing their delicate pink, or white, or even crimson blossoms at different seasons; so that one or other kind may be seen in blossom all the year round. Another portion of the side shelves contains the different plants of the *Pepper kind*, among which will be remarked the true Pepper Plant of our tables (*Piper nigrum*, fig. 44.). and the valuable styptic *Piper angustifolium*, Ruiz and Pav., called *Matico* by the Peruvians, and the *Cubeb* (*Piper Cubeba*).

Due south of the Stove No. 12. we have just been describing, we can, at the northern door, enter a rather small House in the form of a T, a double span-roofed structure;

Nos. XIII., XIV. THE HEATH HOUSE.

No. 13. the *northern wing*, and No. 14. the *transverse portion*.

This is, in part, occupied by a collection of *Cape Heaths* (fig. 45.). Here are several species of the beautiful genus *Epacris*, which

45

CAPE HEATHS.

may be called the Heaths of Australia, being nearly allied to the *Ericeæ*, and perhaps of superior beauty. Even on the outside of this House are some attractive objects.

For example, on the east and west sides of No. 13. are FRAMES containing a miscellaneous collection. That on the east side, in summer, exhibits a noble specimen of *Erythrina laurifolia*, with large coral-coloured papilionaceous flowers; in summer one of the most striking of our half-hardy shrubs. On the west side the frame contains *Cape Bulbs*. On the outside again, in a narrow bed immediately under the *front* of this house (facing the south), with other tender plants, are the three kinds of *Tea*, much cultivated by the Chinese, the *Black Tea* (*Thea Bohea*, fig. 46.), the *Green Tea* (*Thea viridis*), and the *Sasanqua Tea* (*Thea*, or *Camellia*, *Sasanqua*, fig. 47.).* The last

46

BLACK TEA.

47

SASANQUA TEA.

seems only to be grown in China for the sake of the oil; while from the Green and the Black Tea Shrubs of botanists it is generally acknowledged that the Chinese make the green or black tea of commerce indifferently, according to the modes of preparation. In mild winters they may be seen blossoming in the open air so late as Christmas.

Opposite, that is, on the other side of the walk, west of the Heath-house, are two or three interesting hardy trees or shrubs. Sheltered by the old Stove, No. 8., and at its eastern end, stands a fine shrub of the Japanese *Photinia serrulata*, a charming evergreen, seldom bearing flower in this climate. Climbing above it, on the east and north side of the walls of the stove, is a noble plant of the old *Glycine* (now called *Wistaria*) *Sinensis*, whose innumerable clusters of blue flowers

48

SALISBURIA ADIANTIFOLIA.

* These Tea plants have been recently removed to a south wall, facing the *Old* Museum.

(in shape like those of *Laburnum*) are very striking in the early spring, before the leaves are unfolded. Next to that (proceeding north) is a very aged trunk of the singular Japanese "Ginkgo," or *Salisburia adiantifolia* (fig. 48.), whose leaves are shaped like a fan, with a deep notch at the top; and next to that again is the *Terebinth Tree* (*Pistacia Terebinthus*, fig. 49.), considered by some commentators the *El-Elah* of Scripture (generally translated *Oak*): it yields the Scian turpentine, a rare gum, mostly consumed in the Levant.

49

PISTACIA TEREBINTHUS.

It will be desirable to retrace our steps a little if we wish to visit some objects in the portion of ground formerly the Royal Kitchen and Forcing-Grounds, and we shall find a walk on the east side of the Succulent House (No. 7.), which conducts the stranger in an easterly direction into the HERBACEOUS GROUND, or general collection of *hardy exotic herbaceous plants*, including *Grasses*, systematically arranged and intermixed with shrubs and ornamental trees upon the lawn. The same piece of ground contains several Plant-houses, also a building, formerly a fruit-house to the Kitchen-Garden, but converted into a

MUSEUM,

destined to receive specimens of *Fruits* and *Seeds* (dried or preserved in pyroligneous acid, or alcohol), *Gums, Resins, Drugs, Dye-stuffs, Sections of Wood* — in short, all interesting vegetable *products*, particularly those that are useful to mankind, in the *Arts*, in *Medicine*, and in *Domestic Economy;* substances which neither the living plants nor the Hortus Siccus can exhibit. But as there is a separate little book published, a Guide to the Objects in this and the NEW MUSEUM, we shall omit the mention of them here, and proceed to

No. XV. MUSEUM STOVE,

situated at the back, or south side, of the Museum, filled with a choice miscellaneous collection of tropical plants, and climbers trained to the back of the wall and rafters. In this House are several, both of an useful as well as ornamental character. Here are the *Nutmeg* (*Myristica officinalis*, fig. 50. page 50.), which yields both nutmeg and mace, spices, of which the consumption is so great, that, according to Stavorinus, of the former 250,000 lb., and of the latter (mace) 90,000 lb., were sold annually in Europe alone; the *Clove* (*Cary-*

ophyllus aromaticus, fig. 51.), which valued spice is the flower-bud, in shape resembling a nail, whence the Spaniards, who discovered the tree, called it Clava, the French Clou, the English Clove; the best *Caoutchouc Tree* (*Siphonia elastica*), from Pará; the *Upas* or *Poison-Tree of Java* (*Antiaris toxicodendron*), to whose well-authenticated virulence it has been the pleasure of poets and travellers to add many a horrifying imaginary incident*; the famous *Cow Tree*, or Palo de Vaca (*Galactodendron utile* of Humboldt), native of the Caraccas, abounding in a milky juice, which is drawn into gourds by tapping, and given to children as we give them cow's milk; the *Xanthochymus pictorius* of Roxburgh, the fruits of which ripen even with us, and yield, on puncturing, a juice which concretes into one kind of *Gamboge*, the most powerful of drastic medicines, and affording the brightest and best known of yellow colours; the rare *Napoleonea imperialis*, and the very beautiful *Gardenia Stanleyana* from Sierra Leone, of which still finer plants are in the Palm-stove; the singular *Lace Bark Tree* (*Lagetta lintearia*) from Jamaica, whose layers of inner bark (there are as many as the portion of the tree yielding it is years old) resemble, without any artificial preparation, exquisite lace; the *Indian Teak* (*Tectona grandis*), extensively used in ship-building, and the equally useful and much rarer *African Teak*, or *African Oak* (*Oldfieldia Africana*); the celebrated *Gutta Percha* plant (*Isonandra Gutta*), kindly sent by Dr. Oxley, from Singapore; *Bitter Quassia* (*Quassia amara*); the *Vanilla* plant; the

50

NUTMEG.

51

CLOVE.

* In addition to the injurious property of this plant, it is known to grow in low valleys of the island, rendered unwholesome by an excess of carbonic acid gas, which escapes from crevices in the ground. In this atmosphere man and beast who unwarily enter, and birds that attempt to fly over (for the gas is said to reach a height of 18 feet from the ground), fall down dead, and the bottoms of such situations are often strewed with the carcasses of various animals which have perished thus, and not from the effects of the Upas.

Malayan *Mangostan* (*Garcinia Mangostana*, fig. 52.), the rich fruit of which we vainly strive to bring to perfection in our stoves. Here also may be seen the *Bread-Fruit* of the Pacific Isles (*Artocarpus incisa*, fig. 53.).

52

MANATOS.

Perhaps the greatest rarity in this House is the famous *Rice-paper Plant* of the Chinese, from the island of Formosa, with great difficulty procured for us by Sir John Bowring. A Case in the Museum explains the nature of the beautiful and well-known papyraceous substance. Two kinds of *moving plants* are here (on the front shelf) well worthy of notice in the summer months; one the *Humble-Plant, Mimosa*

53

BREAD-FRUIT TREE.

pudica (often, though incorrectly, called the *Sensitive* plant, which is *Mimosa sensitiva*):

> " Weak with nice sense the chaste *Mimosa* stands,
> From each rude touch withdraws her timid hands·
> Oft as light clouds o'erpass the summer glade,
> Alarm'd, she trembles at the moving shade,
> And feels, alive through all her tender form,
> The whisper'd murmurs of the gathering storm,
> Shuts her sweet eyelids to approaching night,
> And hails with freshen'd charms the rising light."

The best way to exhibit the sensitive properties, so called, of the leaves, is to cut off suddenly and cautiously the tip of one of the terminal leaflets, when all the other leaflets on that stalk will close, *a pair at a time*, from above downwards; thence the impulse is continued to the adjoining stalks and to the leaflets from below upwards; and then the whole leaf will fall. The other is the *Desmodium gyrans*, usually called the *Moving Plant*, in Bengal the *Telegraph Plant*. Here the movement is voluntary, so to say, not influenced by touch, only requiring a calm, warm atmosphere. The leaves consist of three leaflets, one large terminal one, and two *small lateral* ones. The latter alone are endowed with this wonderful property. There are some or other of them always in motion, by jerks and in circles, or gyrations, in one direction, so as to return to the same point.

Nos. XVI. & XVII. MISCELLANEOUS TROPICAL HOUSES.

A double stove now occupied by tropical plants, having a bark-pit for rearing certain tropical plants which will not bear too much moisture. Here will probably be seen the *Double-Cocoa-Nut* of the Seychelles (*Lodoicea Sechellarum*, presented by Professor Bojer), *Gutta Percha* (*Isonandra Gutta*); young plants of *Mangostan*, *Xanthochymus*, *Nutmegs*; the rare *medicinal Bark* (*Cinchona Calisaya*), Wedd.; *Mammea Apple* (*Mammea Americana*), *Borassus flabelliformis*. or *Great Fan-Palm* of India, most difficult to rear; the *Cow-Tree* of Humboldt, *Durians*, *Upas*, &c.; and on the rafters *Dipladenias* (*Harrisii* and *speciosa*), *Combretums*, *Hoyas*, &c.

No. XVIII. AZALEA HOUSE.

Besides Azaleas, other miscellaneous plants are placed here in summer and autumn.

No. XIX. AROIDEOUS HOUSE,

so called because it is mainly occupied with this family of plants, many of which are remarkable for their *esculent* properties, combined with a powerful *poison*. The tropics abound with them; and even England yields one kind (our *Arum* or *Wake-Robin*), which, both in appearance and qualities, may be considered a type of the rest. It is characteristic of the family that the juices are poisonous, often eminently so; but those juices being removed by cooking, the foliage and farinaceous tubers become esculent and wholesome:

the former is eaten boiled in India under the name of *Indian Kale*, the latter in various warm countries as a substitute for bread; the *Arum Colocasia*, or *Egyptian Arum* (fig. 54.), for example, in Egypt and the East, and the *Arum esculentum* in the West Indies. Our own *Arum maculatum* is commonly collected in the South of England, especially in Portland Island, and the roots made into pastry or used as *Arrow-root*. *Caladium Seguinum* is the *Dumb Cane* of the West Indies; so called because a small quantity of its juice, dropped upon the tongue, causes that organ to swell violently, and prevents the power of speech. A peculiar aspect is exhibited by all the individuals of the *Aroideæ*, independent of botanical character. Hence the importance of seeing in a garden, plants grouped according to their *natural affinities*, as illustrative of their properties; and this we have attempted to do in many instances in the vast collections of this garden.

54

EGYPTIAN ARUM.

Besides the Aroideous plants, which we have now briefly noticed, we have here several of these *Bromeliaceous* and the *Scitamineous plants*, both containing many useful properties. Of the first of these the

55

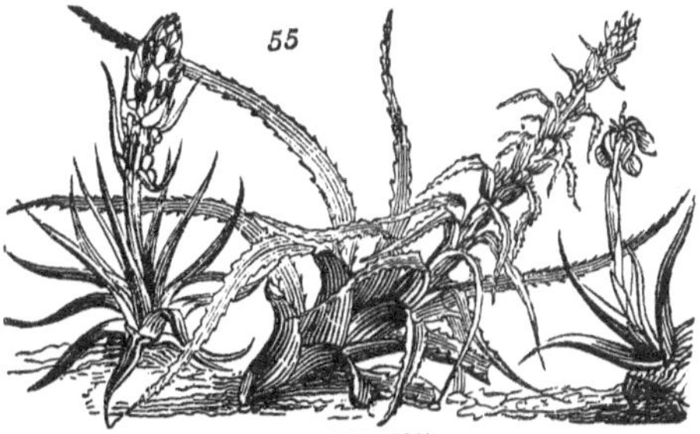

BROMELIA.

Bromeliaceæ or *Pine-Apple family* (fig. 55.), the larger kinds have leaves pre-eminently adapted to the purpose which the plant serves in Mexico and the warm parts of South America, that of making fences. Each leaf, long and sword-shaped, may be seen to have its edges armed with exceedingly sharp hooked spines; those on the upper half of the leaf curve towards the intruder, as if to forbid approach in that direction; while, if he has unfortunately penetrated some way in spite of these innumerable and formidable opponents, the downward curve (towards the centre of the plant) of the remainder of the spines will prevent his egress, except at the sacrifice of the skin and flesh by these strong hooks. The kind which yields the well-known esculent fruit is the *Bromelia Ananas.* Allied to the *true Bromelias* are the *Tillandsias,* which we often receive from tropical countries, and they succeed well with us, attached to truncheons of wood (for they are Epiphytes, like the *Orchideæ*), until the flowering is past, and then they almost invariably wither: we need, therefore, a frequent importation of them. The leaves of many are singularly hollowed at the base, and in the driest weather filled with water, which often proves serviceable to man and animals. Some larger kinds of *Bromeliaceæ* are in the Palm-house, and some in the Aloe-house.

The *Scitamineæ,* contained in this House, include a good many *Spices* and *Medicinal Plants.* Among them we may specify the well-known *Galangale* (*Kæmpferia Galanga,* fig. 56.), the *Arrow-root* (*Maranta arundinacea*), the various kinds of *Indian Shot* (*Canna*), *Phrynium,* and *Hedychium,* with their large and fragrant flowers, *Alpinia, Ginger* (the tuberous roots of *Zinziber officinale*), *Turmeric* and *Zedoary* (both species of *Curcuma*), *Cardamoms* (*Amomum*), &c.

56

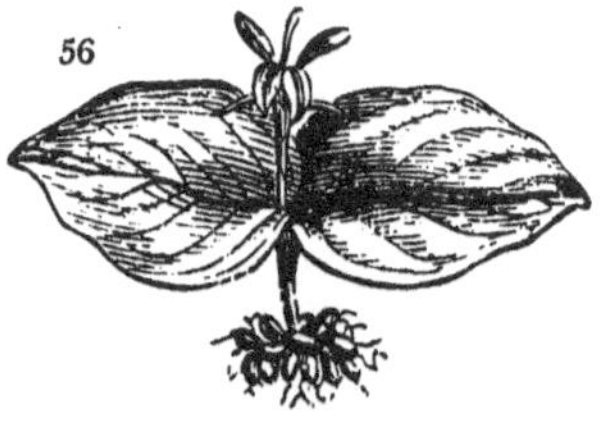

KÆMPFERIA GALANGA.

Some of the smaller kinds of *Screw-Pines* (Pandanus), and fine specimens of the genus *Carludovica,* are here also, the lesser kinds of *Strelitzias,* and some other monocotyledonous plants.

No. XX. TEMPERATE FERNERY,

stored with such Ferns as do not require the tropical heat of those of the Houses Nos. 11, 12. The north aspect of this House is favourable to the growth of these plants, and to such kinds that are natives of temperate climates, Madeira, South Africa, Australia, New Zealand, &c., &c. The House has a few other kinds of plants, among them the rare and handsome-flowered *Lapageria rosea,* from South Chili.

No. XXI. RHODODENDRON AND CAMELLIA HOUSE;

occupied by many tender *Rhododendrons*, including Dr. Hooker's Sikhim Himalayan ones, *Camellias*, *Oranges*, and a very mixed collection of plants of temperate climates, such as tender *Mahonias*, the *green*, *black* and *Assam* Teas, *Paraguay Tea* (Ilex Paraguensis, fig. 57.), no *Tea* indeed, but, as its scientific name implies, a kind of *Holly;* yet, under the name of *Maté*, it affords a beverage almost as extensively used in South America as Bohea, Souchong, or Hyson are in Europe.

57

ILEX PARAGUENSIS.

No. XXII. DOUBLE PROPAGATING HOUSE,

kept private; chiefly occupied by numerous young plants reared from seeds, cuttings for striking, and newly-imported plants.

And now it may be well to return in a westerly direction, past the Houses 11 and 12, where we can enter the Old Arboretum, near a handsome architectural building, *The Temple of the Sun*. This is a nearly circular piece of ground (see the Plan), consisting of about five acres, crowded with hardy trees of much interest and value, more than can possibly be here enumerated. Close by the Temple of the Sun are noble specimens of the *Turkey Oak* (*Quercus Cerris*), a *Cedar of Lebanon* (*Cedrus Libani*, fig. 58.), the *Oriental Plane* (*Platanus orientalis*, fig. 59.), a very large *Locust Tree* (*Robinia Pseudacacia*), a *Lotus* of North America (*Diospyros Virginiana*), a fragrant *Sassafras* (*Laurus Sassafras*), and a healthy though young *Cork Tree* (*Quercus Suber*, fig. 60.), two very large trees of this were blown down in a gale during George III.'s reign; a large *Koelreuteria*, &c. This Arboretum is circumscribed by a walk: by taking that which leads us past the temple, we approach the east end of the

58

CEDAR.

Orangery on the way to the principal entrance, among good exotic trees, ***American Limes, Oaks, Hickories, Red*** **and** ***Yellow-flowered Horse-chestnuts,*** **&c.**

59

ORIENTAL PLANE.

60

CORK TREE.

In the other direction, due north from Stove No. 12., we pass several interesting trees of another description. To the right is a fine *Woolly-fruited Maple* of North America (*Acer eriocarpum*): a little further, and near a private walk leading to the residence of Her Royal Highness the Duchess of Cambridge, and to the Directors' rooms, stands conspicuously a *Weeping Willow* (*Salix Babylonica*), derived from the original tree now destroyed, at Napoleon's tomb, St. Helena; the *Red Maple* of the United States (*Acer rubrum*), the *Manna Ash* (*Fraxinus Ornus*), from which exudes the *Manna* of the shops; the *Glastonbury Thorn*, a variety of the common Thorn (*Cratægus Oxyacantha*), of which the origin was said to be a staff from the Thorn used for crowning our Saviour, which Joseph of Arimathea stuck into the ground at Glastonbury, when it immediately sent forth leaves and flowers! Be that as it may, this tree is remarkable for bearing foliage almost throughout the year, and it flowers, if the season be mild, in winter as well as in spring. There is also a beautiful young tree of the *Deciduous Cypress* of Mexico and the Southern United States (*Taxodium distichum*), which, in its native country, attains to an enormous size, 90 feet in the girth of its trunk, and to a great age: the identical tree at Chapultipec, under which Montezuma was accustomed to sit previously to the conquest of Mexico, is yet living, and known as the Cypress of Montezuma. Here, on the left, are fine old *Celtises*, and the *Paper Birch* of

North America (*Betula papyrifera*). Proceeding towards the entrance gate, we pass the ruin of a stately *Cedar*, of which the main portion was carried away by a gale in 1841, and now completely destroyed (while in a sickly state) by the severe spring of 1854; so that nothing remains but the majestic trunk, which we propose to preserve and clothe with ivy; its age is estimated at about 130 years; and many rare *Pines* from different countries, in various stages of growth. Among them the *Deodar* (*Cedrus Deodara*) ranks pre-eminent, a tree rivalling if not excelling in beauty the *Common Cedar*, and equally hardy, from the mountains of Northern India: this specimen is among the very first reared in Europe, from seeds brought home and presented to the Garden by the late Honourable W. Henry Melville. As the gates of Solomon's Temple at Jerusalem, and those of St. Peter's at Rome, are said to have been made of the *Cedar of Lebanon*, so it has been ascertained, and I believe on sure authority, that the gates of the Temple of Somnauth are constructed of the *Indian Cedar* or *Deodar*. *Abies Webbiana, Smithii, Douglasii*; the *Stone Pine* (*Pinus Pinea*), which is the Pine of Claude Lorraine's Italian landscapes; a very fine Corsican Pine (*P. Laricio*), and the much rarer species, *P. Coulteri, macrocarpa*, and *Sabiniana* from the Pacific side of North America, also find place here.

After the inspection of these, the visitor will find himself returned to the gate on the Green by which he had entered.

INDEX.

THE END.

London: Printed by Spottiswoode & Co., New-street Square.

OLD MUSEUM.

Interior of the Principal Room.

MUSEUM

OF

ECONOMIC BOTANY

OR

A POPULAR GUIDE

TO

THE USEFUL AND REMARKABLE VEGETABLE PRODUCTS
IN THE TWO MUSEUM BUILDINGS

OF THE

ROYAL GARDENS OF KEW.

BY

SIR W. J. HOOKER, K.H.

D.C.L. OXON., F.R., A. & L.S., ETC.

CORRESPONDING MEMBER OF THE IMPERIAL INSTITUTE OF FRANCE,

DIRECTOR.

"Behold, I have given you every Herb bearing seed, and every Tree bearing fruit: to you it shall be for meat," "and for the *service* of Man." *Gen.* i.; *Ps.* civ.

THIRD EDITION.

LONDON
LONGMAN, BROWN, GREEN, LONGMANS, & ROBERTS.
1858

LONDON
PRINTED BY SPOTTISWOODE AND CO.
NEW STREET SQUARE

PREFACE.

The Museum of Economic Botany now occupies two separate buildings, within the Royal Gardens; firstly,

THE OLD MUSEUM;

so called because it was the first structure occupied by objects of this kind; the history and origin of which are as follows:—

In the year 1847, Her Majesty Queen Victoria was graciously pleased to command that those grounds at Kew which had been devoted to the kitchen and forcing department, should be added to the Royal Botanic Garden, together with the stoves and other buildings. An excellent brick structure, partly occupied as a dwelling by one of the foremen, and partly as rooms for preserving fruit for the use of the Palace, was thereby vacated; and it occurred to the Director that, with a little alteration, this might be made a depository for all kinds of useful and curious *Vegetable Products*, which neither the living plants of the Garden nor the specimens in the Herbarium could exhibit; and that such a collection would render great service, not only to the scientific botanist, but to the merchant, the manufacturer, the physician, the chemist, the druggist, the dyer, the carpenter and cabinet-maker, and artisans of every description, who might here find the *raw material* (and, to a certain extent, also the *manufactured* or *prepared article*) employed in their several professions, correctly named, and accompanied by some account of its origin, history, native country, etc., either attached to the specimens or recorded in a popular catalogue. The suggestion was communicated to the heads of the Establishment, the Commissioners of Her Majesty's Woods and Forests: leave was asked to convert the building (at first, partially) into a Museum, and orders were given for one large room to be formed, with a gallery, side-lights and sky-light, and fitted up with glazed mural and table cases. The foundation of the Museum consisted of the Director's private collection (presented by himself), some few objects already belonging to the Garden, chiefly fruits and seeds, and some given by Mr. John Smith, whose son, Mr. Alexander Smith, received the appointment of Curator. Scarcely were the doors opened to the public, when specimens of various kinds were received from all quarters, so that it soon became necessary to enlarge the accommodation, till, in the past year (1855), all the ten rooms of the building and all the passages were completely filled. One had but to see the crowds frequenting in summer the existing Museum (so great that the Director finds it frequently impossible to enter at public hours with distinguished visitors, who desire to have objects explained to them), to be satisfied of the deep interest the public take in such a collection,—a collection that at once appeals to the faculties and understanding, showing the practical uses of the study and application of Botany, and the services thus rendered to mankind. It has done more to recommend and to popularise (if I may use the expression) the science that communicates a knowledge of the vegetable creation, than all the princely Palms, the gorgeous Water-Lilies, the elegant Ferns, &c. &c., which grace the tropical houses of these noble Gardens. Here (in these Gardens) it is true the public are privileged to see growing Plants of 1. the *Cocoa-nut;* 2. the *Vegetable-Ivory Palm;* 3. the *Wax-bearing Palms* of *Brazil* and of the *Andes;* 4. the *Piaçaba Palm;* 5. the *Coco de Mer;* 6. the *Doum Palm;* 7. the *African Oil-Palm;* 8. the plants yielding *Chocolate, Tea, Coffee,* and *Sugar;* 9. the *Papaver somniferum;* 10. the *Bermudian Juniper;* 11. the *Isonandra Gutta;* 12. the *Chinese-grass Plant* (so called); 13. the *Siphonia elastica;* 14. *Rice-paper Plant* (so called) of China, and a hundred others of equal interest: but the visitor receives twofold gratification and twofold instruction, if, with these several vegetable forms in his mind, he can enter an adjacent building, and there contemplate their *products,* and see, as it were, the uses which the ingenuity of man has derived from them:—1. the food and raiment, the milk, the oil, the wine (toddy), the cups and bowls, cordage, brushes, mats, in short, the 365 articles ("as many," the Hindoos say, "as there are days in the year") afforded by the common *Cocoa-nut* alone: 2. the ivory-like articles made from the seeds of the *Vegetable-Ivory Palm:* 3. the excellent and now commercial wax, yielded by certain *American Palms,* and deposited by nature on their trunks and leaves: 4. the useful nuts (for turnery), and still more useful fibre(for the best brooms and brushes) of the *Piaçaba,* as the streets of our great cities can testify: 5. the curious double nuts (with their flowers) of the *Coco de Mer* of the Seychelles Islands, once, while ignorantly supposed to be the product of a submarine tree and only known from having been found floating in the Indian Ocean, valued at the price of a freighted ship, being deemed an antidote to all poisons, "mirum miraculum naturæ," says Rumphius, "quod princeps est omnium marinarum rerum, quæ raræ habentur:" 6. the singularly forked stems (other Palms have unbranched stems), and the large ginger-bread-tasted fruits of the *Doum-Palm* of Upper Egypt: 7. the fruits and nuts of *Elais Guineensis*, which yield the Palm Oil of western tropical Africa, now largely imported and consumed by Messrs. Price and Co.; and not those only, but the oil itself and the several preparations it undergoes in its progress towards the perfect candle, in a series of samples presented by that truly philanthropic company: 8. the various and generally well-known preparations of *Chocolate, Tea, Coffee,* and *Sugar*, the latter extracted from an *American Maple* and from *Beet-root*, as well as from the *Cane:* 9. *Opium* in its various stages, from the "*poppy-head*," to the pernicious ball prepared for commerce: 10. the progress of pencil-making of the wood of a *Juniper*, improperly called *Cedar:* 11. the various interesting products derived from the *Gutta-Percha plant:* 12. the fibre or raw material and cloth, made from the *Chinese* "*Grass plant*," in reality a *Nettle:* 13. cases filled with the various preparations of the *best Caoutchouc* or India-rubber tree of Pará (*Siphonia elastica*), presented by Messrs. Macintosh: 14. the beautiful substance called Chinese *Rice-paper*, long supposed to be a preparation of rice, but here shown to be the exquisite pith of a new plant (*Aralia papyri, era*), only found in Formosa, for the knowledge and for living plants of which we are mainly indebted to Sir John Bowring. The last-mentioned discovery is only one of many instances (as will by and by be shown) of the *origin* and history of commercial vegetable products, of which we should yet have remained in utter ignorance, but for the formation of this Garden and this Museum: such information cannot fail to be acknowledged and appreciated in a great mercantile country by every intelligent mind, and even by persons not specially interested in trade or commerce.

We flatter ourselves, too, that the general arrangement of the collection has proved instructive; and the marked attention of many visitors, the notes that are taken in the room, and the works of artists performed there, show that it is so.

While the collection was comparatively in its infancy, we adopted what may be called a commercial arrangement of the articles, bringing them together as much as possible under the heads of 1. *Fibres* or *Textiles;* 2. *Gums* and *Resins;* 3. *Dye-stuffs;* 4. *Starches;* 5. *Oils;* 6. *Woods;* 7. *Tannins;* 8. *Drugs;* 9. *Food for man;* 10. *Basketwork*, &c. &c.; but this plan, though suitable in many respects, was attended with manifest inconvenience. Even to a botanist it conveyed no idea of the *kinds* of plants yielding such and such substances, or possessing such and such properties; and it had this further difficulty, that, to make each series complete, several objects had to be *repeated* under two or more different heads, many drugs being used as food, while some dye-stuffs are valuable as woods, &c. It appeared more advantageous to arrange the collection according to the natural affinities of the plants producing them; and, with the exception of many woods (of which the names are yet unknown), and of some mixed substances (*Peats*, for instance), or such as from their size or other causes will not rank in their proper places, the arrangement is that followed by Professor Decandolle, in his "System of the Vegetable Kingdom," commencing with—1. The Dicotyledons or Exogenous Plants (headed by the *Ranunculus family*): 2. The Monocotyledons or Endogenous Plants, to which the *Grasses, Palms*, &c., belong: and concluding with the Acotyledons or Cryptogamic Plants, to which are referred *Ferns, Mosses, Lichens, Fungi, Algæ;* these, however numerous and however curious to the botanist, ministering comparatively little to the immediate wants of mankind.

An early part of the following Catalogue shows, for example, that the *Ranunculus* or *Crowfoot-family* contains acrid and poisonous principles: the *Poppy-family* is narcotic, while the seeds are wholesome: the *Sour-sop-family* supplies excellent fruits: the *Cruciferæ* are antiscorbutic, and the *Violet-family* is emetic; while more familiar instances may be mentioned in the *Grass-family*, which yields all our cereal grains and abundant fodder for cattle: the *Labiate-family* is aromatic and contains volatile oil: the *Bindweed-family* is cathartic: the *Gentian-family* abounds in bitter principle, the *Conifers* in Resin, Turpentine, and Tar, the *Mallow-family* in mucilage and fibre. Possessed of this kind of knowledge (I mean, an acquaintance with the natural families of the Vegetable Kingdom), an intelligent traveller may safely estimate the properties which a plant, though he has never seen it before, may possess; and many instances are narrated of botanists, who have accompanied exploring expeditions and voyages of discovery, unhesitatingly employing certain plants hitherto unknown, as food, or for the relief of some malady, especially the scurvy. If in want of string or cordage, the bark of the *Mallow-* and *Nettle-family*, and equally of the *Spurge-laurels*, &c. &c., would afford it. He would be aware too, that many plants, justly deemed poisonous in the crude or raw condition, possess this principle in a volatile state, as for example the *Potato* (though not to a high degree), the *Arum-family*, and the *Cassava* or *Tapioca* plant (eminently poisonous); yet, when the injurious principle is dissipated by cooking or forced out by pressure (witness the tapioca-bags of the South American Indians in the Museum), the residue affords nutritious food, the staff of life to millions.

The utility of this Museum to mankind at large, and of that arrangement of the collection which we have been discussing, is testified by the remark of not a few visitors, "Now we see, for the first time in our lives, and on a large scale, a *practical* application of the *science of Botany.*"

In the publication of the following Catalogue and notes, we avow another object, besides giving instruction, viz. that of showing our deficiencies, and affording our readers the opportunity, which many will gladly embrace, of contributing numerous objects not here recorded, but which it is obvious will be prized and gratefully accepted in a Museum destined, in the most public and efficient manner, to exhibit all *curious vegetable products,* and all which minister to the *wants and to the convenience of Man.*

Since the date of the first edition of this Guide, in the autumn of 1855, the following Report on the Old Museum, which formed a part of that on the Royal Gardens, was prepared and laid before Parliament by the Director:—

"This Building afforded 6,000 square feet of glazed cabinets, for the deposit of those vegetable products that are either eminently curious or in anywise serviceable to mankind; it is consequently of the highest importance, if simply for instructional purposes, and essential to a great commercial country, as must be evident to any person who reflects on the innumerable ways in which the Vegetable Kingdom ministers to the wants of the human race. The ship-builder, the carpenter, the cabinet-maker, the general merchant, the manufacturer, the weaver, the physician, the druggist, the dyer, the oil and colourman, &c. &c., will here find, systematically arranged, the several objects in which they are interested, accompanied by their correct appellation, the countries whence they come, and the names of the Plants from which they are procured. Till this Collection, and the correspondence which it entailed, determined these points, many had been entirely or chiefly unknown as regarded their origin or native country; and it is indubitable that every year more and more information will be elicited on points of no slight importance to a great mercantile country. So valuable and extensive a Collection required a cheap and popular Guide Book, similar in size and plan to that for the Gardens; and the present has consequently been published.

"This Museum had already become, even before the close of 1854, inconveniently crowded by the accession of new contributions, and a considerable amount of them had to be temporarily accommodated in the Temples and in Sheds widely apart from the Museum building, and at present not available for public use. A Vote for an additional Museum was, therefore, by our Chief Commissioner, submitted to, and sanctioned by, Parliament,"—and the result is the erection of

THE NEW MUSEUM,

opposite to the Palm House, the ornamental water intervening, a structure destined to give further accommodation to the extent of 13,000 square feet of mural glazed cabinets. And this was of the greater importance, for it was in the summer of 1855 that the Director was summoned, by the French Imperial Commission, to take an active part in the Jury of the Second Class at the Great Paris Exhibition. That class, mainly relating to Vegetable Products, offered the very objects most requisite to enrich our collection. A grant of 200*l*. was accordingly obtained from the Treasury, to be spent in procuring such articles as were most important for the Kew Museum, while the President of the Board of Trade liberally offered an equal sum. Thus provided, and further assisted by the several Officers of the Science and Art Department of the Board of Trade, then in Paris, and enriched by numerous donations from many Exhibitors, 48 large cases were transmitted to Kew. They contained Vegetable Products, many of them very rare and valuable, from Algeria, Australia, Austria, the East Indies, France, the Grand Duchy of Hesse, Jamaica, Mauritius, Norway, Prussia, Sardinia, Sweden, Tuscany, the United States, Tasmania, Victoria, Wurtemberg, &c. &c. All have arrived in safety, and their contents will go far towards filling up the surplus space in the new Museum.

To say that this Collection of Vegetable Products is unrivalled, is saying little, and no more than might have been asserted while the Museum was quite in its infancy, since nothing of the same instructive kind had ever been attempted. Ours is the gratification of having set the example, which is now being followed in several of our colonies (Jamaica, Demerara, Melbourne, &c.). The East India Company is forming a similar Museum in London, at Calcutta, and at Madras; another has been attached to the Botanic Garden of Edinburgh, Liverpool, Manchester, &c. A small, but well arranged, series of similar objects was formed towards the close of the Paris Exhibition, by order of his Imperial Highness the Prince Napoleon; and we are likely to have a powerful but honourable rival in the Museum of the Jardin des Plantes at Paris. There, indeed, an unexampled Collection of Woods, and other products connected with

Physiology and pure Botany, have long been deposited, but with no approach to an Economic Museum till recently, when donations and purchases, made at the close of the Great Paris Exhibition, laid the foundation of one there on a most extensive scale.

Already our Museum owes much to the donations of friends whose names accompany the respective articles presented by them; and we cannot here specify more than the "Lawsonian Collection," (which formed a part of the Great Exhibition of 1851, presented by Messrs. Peter Lawson and Son, and on which they published a volume entitled "The Lawsonian Collection, or Synopsis of Vegetable Products of Scotland in the Museum of the Royal Gardens of Kew, 1852");—the various Indian products procured by Dr. Hooker during his Mission to Sikkim Himalaya; the produce of the Rice-paper plant from China from Mr. Bowring; raw material and prepared articles of Gutta Percha from the Gutta Percha Company; of Caoutchouc from the Company so named; of vegetable waxes, tallows, and oils from Messrs. Price and Co.; a fine collection of preserved fruits and other vegetable substances from Messrs. Fortnum and Mason, &c. &c. We owe much to the public Government-offices of this country, for their powerful aid in various ways, especially to the Admiralty and the Foreign and Colonial Departments. James Wetherell, Esq., late H. B. M. Consul-General at Bahia, has sent many valuable collections from that country; and I must also record the name of the Rev. Professor Henslow, who not only contributed various preparations of great interest and importance, but gave much advice and assistance in the arrangement of the Museum. We proceed now to notice the contents of the two Museums, beginning with the New Museum, No. 1.

Royal Gardens, Kew: June 1, 1858.

CATALOGUE,

WITH REMARKS,

OF THE MORE INTERESTING OBJECTS IN THE MUSEUM SYSTEMATICALLY ARRANGED.

NEW MUSEUM BUILDING, No. 1.

CONTAINING

Cl. I. DICOTYLEDONS or EXOGENS.

Second or Top Floor (including Cases 1—36).*

Ord. Ranunculaceæ. Crowfoot Family. (*Case* 1.) — The *Crowfoots*, *Aconites*, and *Larkspurs*, &c., are familiar examples of this Natural Family, and remarkable for the virulent poison which they often contain ; so that, though few or none are articles of food, a great many yield powerful medicines. They are herbaceous, with annual or perennial roots. Acrimony, causticity, and poison are said to be the general characters of this Order, often combined with narcotic. Notwithstanding, however, these potent properties, it is remarkable how few *Ranunculaceæ have been*, or are still, *generally* employed in medical practice. The Hellebore is the chief medicine that is retained, and, from its uncertainty, is but seldom used.—*Hepatica.* Hepatica triloba. Eur. and N. Am. Astringent.—*Pasque-flower.* Anemone Pulsatilla. Eur. Leaves. Recommended in cutaneous diseases, &c.— *Crowfoot, upright-meadow.* Ranunculus acris. Roots and leaves. —*Lesser Spear-wort.* Ranunculus Flammula. Eur. Chiefly employed in rustic practice, being acrid and irritant; taken internally they occasion vomiting. — *Hellebore root*, and *Extract.* Helleborus niger, Often called *Christmas Rose.* S. Eur. Drastic purgative and diuretic. Largely imported from Hamburg, and sometimes from Marseilles.— *Fetid Hellebore.* Helleborus fœtidus. Green leaves; chiefly used in rustic practice, and Extract as anthelmintic. N. Am.— *Gold Thread.* Coptis trifolia. U. S. Pure and powerful bitter.—*Mishmee Teeta.* Coptis Teeta. Assam. Tonic.—*Fennel-flower.* Nigella sativa, Eur. and N. India. Seeds. Formerly employed instead of pepper, being pungent: still used in India as a condiment and a medicine.—

* The cases are all numbered on the top at the end and in the front.

Stavesacre. Seeds. Delphinium Staphisagria. Extremely poisonous, a property due to a peculiar alkaloid called *Delphinia.* Emetic, drastic, and inflammatory; chiefly used in external applications. S. of Eur.—*Field Larkspur.* Delphinium consolida. Seeds. Yields a blue dye.—*Aconite Root* and *Extract.* Aconitum Napellus. Eur. Narcotico-acrid poison. The fact of *Aconite* roots having been carelessly mistaken for *Horse-radish,* and thus occasioning the death of several persons, has induced us to place the roots of each in this case, with remarks on their distinguishing characters. Sample of pure *Aconitine; Mannite* from Monk's-hood root.—*Bikh.* Aconitum ferox. Nepal and Sikkim. Root. Amongst the most virulent of vegetable poisons. The Goorkahs employ it to poison their waters.—*Butees.* Aconitum heterophyllum. Simla. Used as a tonic in India.—*Pæony.* Pæonia officinalis. Eur. Flower-seeds and root: supposed to be antispasmodic; the latter emetic and cathartic.—*Bane-berry.* Actæa spicata. Eur. Berries poisonous. Roots antispasmodic and astringent. — *Black Cohosh Bark.* Cimicifuga racemosa. N. Am. Nauseous and astringent, reputed to be a remedy for the bite of the Rattle-snake. — *Mandrake-root, American* or *May-apple.* Podophyllum peltatum. N. Am. Cathartic. Leaves poisonous; the fruit acid, hence called *Wild Lemon.*—*Yellow-root.* Xanthorhiza apiifolia. N. Am. Shrub contains both gum and resin, intensely bitter.

Ord. Magnoliaceæ. Magnolia Family. (*Case* 1.)—The well-known *Magnolias* are the type of this Natural Family, which, however conspicuous by luxuriant foliage, large and beautiful and highly fragrant flowers, does not yield much to economic botany. The bark is considered tonic and aromatic, and the fruits also.—*Swamp-sassafras,* or *Beaver-tree.* Magnolia glauca. Fruit. Both the fruit and the bark in the U. S., where it is a native, are in high esteem for their bitter and aromatic principle, rivalling, it is said, Cinchona.—*Champaca.* Michelia Champaca, in liquid. China. All the parts of the tree highly stimulant.—*Winter's bark.* Drimys Winteri. Continent of S. Am. Wood and bark. Stimulant, aromatic, and tonic. — *Star-Anise.* Illicium anisatum. Fruit. Japan. Aromatic and carminative. The oil is often substituted for that of common Anise.

Ord. Anonaceæ. Custard-apple Family. (*Case* 1.)—Mostly tropical trees and shrubs. The fruits of many are esculent, and much prized; properties aromatic.—*Soursop,* or *Rough Custard-apple.* Anona muricata. Fruit. W. Ind. and S. Am. Eaten, often mixed with sugar, and wine, and nutmeg, being a grateful acid. Cultivated in Western Africa, whence a necklace, formed of the seeds of this fruit, has been sent to the Museum.—*Sweetsop,* or *Scaly-fruited Custard-apple,* Anona squamosa. (*Fig.* 1.) Native, it would appear, of the Malay Islands, but abundantly cultivated in the E. and W. Indies. Much esteemed at table by some, spoken lightly of by others. The Creoles are exceedingly fond of it, but Dr. M'Fadyen says he never knew a European who liked it.—The *Cherimoya* or *Cherimolia* of Peru, Anona Cherimolia. Cultivated in Peru. Judging from the figure of Feuillée, Per. vol. iii. p. 17., it must be very nearly allied to *A. squamosa.* The fruit is, like it, scaly, and vaunted by the Creoles as the most delicious in the world; while Father Feuillée remarks, and probably other Europeans would agree with him, "il est certain qu'une de nos Poires ou de nos Prunes valent mieux que toutes les Cherimollos du Pérou."—*Common,* or *Netted Custard-apple, Bullock's heart.* Anona reticulata. Fruit eatable. Said to be a native of the W. Indies, and thence introduced into the E. Indies and the Malay Archipelago.—*Negro-pepper,* or *Guinea-pepper.* Habzelia Æthiopica. Aromatic and pungent; the *Piper Æthiopicum* of old authors; used as pepper by the natives of the African coast.—Oil from the flowers of Uvaria odorata. Amboyna. —*Frutto de Burro.* Xylopia longifolia. Venezuela.—*Bitter-Wood bark.* Xylopia glabra. Jamaica. Taste like orange-seed. Fruit agreeable, giving a delicate bitterish flavour to the wild pigeons that feed upon them.—*Yari-Yari bark.* Duguetia Guitarensis. B. Guiana. "Bark stimulant and sudorific, possessing scarcely any astringent properties;" said to afford the *Lancewood.*

Fig. 1.

ANONA SQUAMOSA.

Ord. MENISPERMACEÆ. MOON-SEED FAMILY. (*Case 2.*)—This Natural Family is named from the lunate form of the seed and especially of the embryo, and consists chiefly of tough climbing shrubs, the wood often without zones or concentric layers. They possess powerfully bitter and narcotic qualities, and are nearly all tropical, and chiefly Asiatic, plants.—*Calumba* or *Kalumb root.* Jateorrhiza palmata. "One of the most useful stomachics and tonics." Mozambique. The large thick roots are dug up by the natives, and the offsets are sliced, strung on cords, and hung up to dry in the shade.—*Cocculus Indicus,* or *Indian berries.* Anamirta Cocculus. India. Fruits. A highly poisonous drug, extensively employed for intoxicating fish, and by unprincipled brewers in the manufacture of beer. Imported sometimes to the extent of 2359 bags, each of one cwt., annually.—*Pareira brava,* or *Velvet-leaf.* Cissampelos Pareira. Root. S. Am. Possesses a tonic and occasionally diuretic power.—*False Calumba root* (or rather *stem*). Coscinium fenestratum. Ceylon. A root of some interest, on account of the frauds that have been attempted with it, large quantities having been sent to England from

Ceylon, as *true Calumba-root.* The internal organisation of this stem is very curious.

Ord. NYMPHÆACEÆ. WATER-LILY FAMILY. (*Case* 2.)—The large rhizomes, or underground stems, of these aquatics, as well as the seeds, are esculent, yet rarely used by civilised man. In some countries the former (commonly called the roots) are used for feeding swine. But if this family be deficient in practical utility, it holds the first rank for the size and beauty, and frequently fragrance, of its flowers, and the amplitude of its foliage.—*Victoria Water-Lily.* Victoria regia. Native of many of the still waters of tropical America, east of the Andes. *Flowers* and young *leaf.* A series of four large drawings, exhibiting the botanical characters of this plant, is on the wall of the ground floor.—*Prickly Euryale.* Euryale ferox. E. Indies. Seeds eaten by the Hindoos. The plant may be considered a miniature *Victoria* with leaves no larger than the White Water-Lily, and flowers smaller than any of the Family.—*Downy Water-Lily.* Nymphæa pubescens. E. Ind. Root. Much eaten in Scinde.—*White Water-Lily.* Nymphæa alba *L.* Eur. Roots, used in Scotland for dyeing black or gray, and in France for tanning leather, and preparing a sort of beer.—*Yellow Water-Lily.* Nuphar lutea. Eur. Fruits, in alcohol. The leaves are said to be styptic. The flowers have a brandy-like smell, and the pistil is shaped like a flask, whence the name of "Brandy-bottle."

Ord. NELUMBIACEÆ. WATER-BEAN FAMILY. (*Case* 2.)—A Family of aquatic plants, confined to two species, one a native of the tropical and subtropical portions of the Old World, and one of like regions in the New World. The former is of great classical interest, *Nelumbium speciosum,* as considered to be the *Egyptian Bean* (κυαμος) of Pythagoras. It was formerly common in Egypt as well as India, but is now extinct in the former country. The seed-vessel, or receptacle of the fruit, is large, in shape an inverted cone, and has the nuts placed loose in apertures or cells on the surface: it has been not inaptly compared to the rose of a watering-pot. The whole, Sir James Smith tells us, "in process of time separates from the stalk, and, laden with ripe oval nuts, floats down the water. The nuts vegetating, it becomes a cornucopia of young sprouting plants, which at length break loose from their confinement, and take root in the mud."—*Sacred Indian Lotus,* or *Bean of Egypt.* Nelumbium speciosum. India. A group of flowers and foliage admirably modelled in wax, (on the table of the room No. 1, Old Museum). *Fruit* dry, and *necklaces* of the nuts. *Root-stock* from Scinde, where and elsewhere it is much eaten boiled and in curries, raw or roasted. Necklaces made of the nuts are sold at Benares, and considered very sacred. A drawing of the entire plant is suspended upon the wall of the stairs.—*Yellow Nelumbium*, or *Water Chinquepin*. Nelumbium luteum. *Flowers* in fluid, and ripe *capsules* with *nuts*. Trop. Amer. and S. U. S. Root-stocks resemble those of the Sweet Potato, and, when boiled, are farinaceous and agreeable; and employed as food by the Osage and other Western Indians.

Ord. BERBERIDEÆ. BERBERRY FAMILY. (*Case* 2.)—Acidity of fruit, from the presence of malic acid, and yellow dye afforded by the wood and root, exist probably in all the *Berberries*. The fruit of *Berberis aristata* is said by Dr. Royle to be dried by the Hill-men (Himalaya) in India, as Raisins are prepared. The curious irritability of the stamens is known to most observers of nature. — *Common Berberry*. Berberis vulgaris. Europe. Wood and bark and root yield a yellow dye, and the root is employed in the manufacture of Morocco-leather. The berries make an excellent preserve. —Woods of various species are exhibited.

Ord. SARRACENIACEÆ. SIDESADDLE-FLOWER FAMILY. (*Case* 2.) —Though no peculiar properties are known to exist in this Family, its remarkably formed foliage cannot fail to attract attention.—The leaves in the Museum are those of the *Red Sidesaddle-flower*, Sarracenia rubra; *Purple Sidesaddle-flower*, Sarracenia purpurea, (drawing on this floor); and *Parrot's beak Sidesaddle-flower*, Sarracenia psittacina. All from N. Am.

Ord. PAPAVERACEÆ. POPPY FAMILY. (*Case* 3.)—This possesses narcotic and acrid properties in an eminent degree; the juice is often white, yellow in *Chelidonium majus*, red in *Sanguinaria Canadensis*. One species alone, the OPIUM POPPY may, in its legitimate use, be reckoned amongst the greatest blessings to mankind, "Magnum Dei donum;" and by its misuse, the greatest curse. Pereira's "Elements of Materia Medica" bear ample testimony to the former; and the "Confessions of an Opium-Eater" to the latter. 6,500,000 lb. are said to be annually bought up for manufacture and as a source of revenue by the E. I. Company. Little more than 100,000 lb. is required for the use of this country, but it is calculated that 20,000,000 lb. are annually consumed by mankind, representing a value of 20,000,000*l*. in money. Our present business is mainly with the contents of our Museum, as bearing on this Natural Order. The seeds in this family are oily, and generally not narcotic.

White, or *Opium Poppy*. Papaver somniferum. Asia and Egypt. Extensively cultivated in warm climates. Through Dr. Hooker's Mission to India, a most valuable collection, elucidating the preparation of Opium, from the Poppy-heads, or green capsules of this plant, as prepared at Patna, is presented by Dr. Corbett of that establishment to the Museum. Taken in detail, this series may be said to commence with a number of small drawings in one frame on the outside of Case No. 10, exhibiting the mode of cultivation, &c. of the Opium; viz., *Fig.* 1. represents the ploughing of the field preparatory to the sowing of the seed. 2. Breaking the clods. 3. Levelling the ground. 4. Irrigating. 5. Laying out the field in compartments. 6. Sowing the seed. 7. Weeding. 8. Watering the plants. 9. Gathering the "leaves" or petals, for "caking." 10. Preparing the petals in a plate, on a gentle fire. 11. Puncturing the Poppy heads, with the Nashtur (called Mahurnee at Patna), or lancet. 12. Gathering Opium, which flows after the incisions, from the green heads, with the Sittooas or scoops. 13. Placing fresh-gathered Opium in a sloping shallow vessel, to allow the moisture to drain off. 14. Conveying Opium to the sub-deputy agent for weighing.

For the various processes of the extraction of the juice and the preparation of the drug we must refer to the sixth volume of the "Kew Garden Miscellany," contenting ourselves here with a catalogue of the articles deposited in this case, and used for the purposes just mentioned. 1. Some unpatched Poppy pods. 2. Five Mahurnees, or instruments for patching. 3. Some patched Poppy pods. 4. Two sittooas, or scoops, for collecting the Poppy juice from the pods. 5. A brass thallee, or edged plate, in which the fresh Opium is collected. 6. Earthen kurraces. 7. One empty jar, in which the Opium is carried to the Godowns. 8. Bamboo scoop, for taking out Opium from jars for examination. 9. A rake for mixing the Opium in the vat, and making it of one uniform quality. 10. A round tin tagar, or pan, used for holding samples, with its cover. 11. A kareegur, or cake-maker's seat. 12. A tin tagar, for holding Opium sufficient for five cakes. 13. A stool for keeping tagars. 14. Poppy leaves, used for wrapping the drug; sufficient for one cake. 15. Some *Puttle*, or petals (not *Puttee*, which is used for filling up the interstices of the chest compartments: *vide* No. 24.), used for pasting and repairing the cakes. 16. A tray, in which Poppy leaves are served out to the cakemakers. 17. A brass cup, used in caking. 18. A shell, half-made. 19. A brass lewah cup. (This cup contains lewah — a paste made of bad opium, used for cementing the petals — sufficient for one cake.) 20. A tray, in which the Opium balls are made. 21. Earthen cups, in which the balls are dried. 22. A ball of Opium, ready for the China market, equal in value to 1 seer 10 chittacks of Opium, at 75° consistence factory weight, and $5\frac{1}{2}$ chittacks of lewah or paste at 51° consistence, and $5\frac{1}{2}$ chittacks of Poppy leaves, or puttle. 23. A chest, with its compartments, ready for filling. 24. Some Poppy trash, or Puttee, used in filling up the interstices of the compartments in the chest. 25. A cake of Opium, as prepared for the Medical Board. 26. A small specimen of Opium, adulterated by water, in a bottle. 27. Opium, adulterated by pussewah, in a bottle.

We find, in further connection with Opium, —*Poppy-heads*, varying in form and size, from different localities. From these also, in a dried state, the decoction, syrup, and extract of Poppies are prepared.—*Poppy-heads*, which have been attacked by tomtits (Parus cæruleus) for the purpose of getting out the seeds; making almost a skeleton of the capsules.—*Seeds of* Papaver somniferum. These are not narcotic, but yield a bland oil, "similar to that obtained from Olives." Samples of the *Oil.*—*Turkey* or *Smyrna*, and *Egyptian Opium, Persian Opium*, from Trebizond; and *English Opium.*—*Poppy petals.* Petals of the *Corn Poppy* (Papaver Rhœas). Eur. Employed in the preparation of the syrup; but chiefly used as colouring matter.—*Argemone Mexicana.* Tropical countries. Fruit and seeds. Infusion occasionally used medicinally. *Oil* of ditto, called *Brumadundoo* oil. Tanjore.—*Blood-root.* Sanguinaria Canadensis. U. S. and Canada. Root and extract. Acrid narcotic.—*Celandine.* Chelidonium majus. Eur. Extract. Acrid poison; yet used medicinally.

Ord. Cruciferæ. Cruciferous Family. (*Case* 4.)—Derives its name from the cross-shaped petals, there being almost invariably four uniform petals to the flower, placed in opposite pairs, hence cross-shaped; and the six stamens (four long and two short) afford almost as constant a character. Herbaceous, abounding in genera and species, chiefly inhabiting temperate climates: none of them is poisonous or hurtful; a great number are useful to mankind, especially the less acrid, as esculent, culinary; affording oil in the seed, and oil-cake for feeding cattle, and valuable manures. Many are antiscorbutic, particularly the *Cochleariæ* (Scurvy-grasses); but such plants become inert when dried. They contain sulphur and nitrogen, to which is supposed to be due their animal odour when rotting. As this Family of Plants is rich in culinary and horticultural and agricultural objects, not a few of them owing their peculiarities to cultivation, and as our extensive collection of them is the Messrs. Lawson's liberal gift, already mentioned in the Preface, I must here refer to their excellent catalogue for what concerns the agricultural products of this Family.—*Water-cress.* Nasturtium officinale. Brit. Seed. Young plants are a favourite salad, pungent and antiscorbutic.—*Cuckoo flowers.* Cardamine pratensis. Brit. Medicinal: considered useful as a stimulant, diaphoretic, and diuretic.—*Horse-radish.* Cochlearia Armoracia. Eur. Seed, and wax model of root. Pungent, acrid, stimulant, and vesicant. Scraped roots much used as a condiment.—*Rose of Jericho.* Anastatica Hierochuntica. (*Fig.* 2.) Syria. A small plant, growing in exposed deserts, where it is often uprooted and blown about by the winds; it has the property of rolling up like a ball in dry weather, opening and spreading out its branches in wet weather. The most absurd fables of the virtues of this plant are related in the East, and greedily believed. It is the original "Rose of Jericho."—*Garden-cress.* Lepidium sativum. Eur. Seed, and oil. The young plants, with those of *Mustard,* are frequently eaten as "mustard and cress." Wholesome and antiscorbutic.—*Gold of Pleasure.* Camelina sativa. Eur. Seeds and oil. Much cultivated for the oil throughout Eur. Brooms are made from the dry haulm.—*Woad.* Isatis tinctoria. Seed, and samples prepared for dyeing blue.—CABBAGE TRIBE. Cabbage, Rape, Turnep (*Brassica*).—*Common* or *Wild Cabbage.* Brassica oleracea. Drawings and models. Native of the sea-coasts of England and the middle and south of Eur. This is considered to be the origin whence the numerous cultivated varieties of *Cabbage* have sprung. A large proportion of these are very unsuited to a Museum, from the difficulty or impossibility of preserving such succulent products, and they would not be very instructive. The seeds, correctly named, several models in wax, and drawings, are exhibited, and we can only notice some of the better-defined kinds.—*Tree* or *Cow Cabbage. Seed* and *stems,* and *walking-sticks.* One of the most remarkable of the Cabbage kind, having a hard and woody stalk, averaging, Messrs. Lawson say, five feet high. No one can visit Jersey without being struck with this plant in the kitchen-gardens, not five feet high only, but frequently eight or ten feet; we remember a dried stalk in Mr. Lambert's possession measuring thirteen feet, and Mr. S. Curtis saw one eighteen feet long, in which a magpie had built its nest! This is mainly produced by daily pulling off the lower leaves as fodder for the cows, leaving foliage only at the top; thus a small garden of them has almost the appearance of a little plantation of Palms. Planted close, as living fences, they keep out fowls and small animals. Sheds are thatched with the dried stems; they serve as stakes for kidney-beans, peas, &c., and "as cross-spars" for the purpose of upholding the thatch or roof of the smaller classes of farm-buildings, cottages, &c., and when kept dry, are said to last upwards of half a century. At a distance from the coast, and in colder latitudes than Jersey, this Cabbage always degenerates. Some of our dried stalks measure nine feet in length. The walking-sticks are almost handsome.—Of the other varieties of Cabbage we reckon,

Fig. 2.

ANASTATICA HIEROCHUNTICA.

either in the shape of seeds, or drawings, or models, forty-one kinds, including *Broccoli, Cauliflower, Kohl Rabi,* &c.; many are probably hybrids, though of great importance for agricultural and culinary purposes.—*Turnep.* Brassica Rapa. Eur. This includes all the varieties of the Common Turnep, of which forty-four sorts are represented by seed, drawings, or models, in the Museum. That state of the root called "*Anbury,*" or "*fingers and toes*" (wax model) is well known to farmers, and is a disease supposed to be caused by the soil.—*Swedish Turnep,* or *Common Wild Navet.* Brassica campestris. Eight varieties are in the Collection, and drawings of several. Oil is extracted from the variety called *oleifera.*—*Common* or *Winter Rape, Cole-seed.* Brassica Napus. Seed and oil (common and refined), and Rape-cake. The oil is commonly known as *Rape oil,* and the crushed residuum called Rape-cake is much used for fattening cattle.—*Mustard. Seed, oil, prepared mustards,* with the *bran* or *husks,* and *lawn-sieve* used in separating the mustard used at table from the bran. Two species are in general use in England; *Sinapis alba* (*a. Fig.* 3.), yielding the white or Essex mustard, and *Sinapis nigra* (*b. Fig.* 3.). Cambridge brown mustard. In India, oil is extracted from *S. Chinensis* and *S. glauca.* — *Sea-kale.* Crambe maritima. Eur. Seeds, and wax model. A well-known and delicate vegetable, much improved by the skill of the gardener.—*Crambe Kotschyana.* Scinde. Roots of this we have received as an esculent; as Crambe Tartarica is considered in Hungary.—*Radish.* Raphanus sativus. Of this familiar root there are many varieties, if not hybrids; models and drawings and seed, and a gigantic var. from Portugal. *Oil* from the seed.

Fig. 3.

a. SINAPIS ALBA.
b. S. NIGRA.

Ord. CAPPARIDEÆ. CAPER FAMILY. (*Case* 5.)—A family of no great importance; properties considered similar to those of *Cruciferæ:* many are arborescent; *Capparis excelsa* of Madagascar is said to afford planks four feet broad; few kinds are employed in Europe, except *Capers.* They are the *flower-buds* of a suffruticose plant, common in the South of Europe on rocks and walls, Capparis spinosa. In the warm parts of France and in Italy the shrub is much cultivated. The flower-buds are gathered, and preserved in vinegar, and they constitute a very considerable article of trade.—*Wood* of *Capparis grandis*. Madras. *Bark of common Caper*, Capparis spinosa; used medicinally. France. — *Fruits* of other and unknown *Capparideæ* in the collection are only of botanical interest.

Ord. RESEDACEÆ. MIGNONETTE FAMILY. (*Case* 5.) — Few are unacquainted with the general appearance and curious structure of the flowers of the common Mignonette: it is a native of warmer climates than ours; but we possess in England, and it is common throughout Europe, the *Yellow-weed*, or *Weld*. Reseda Luteola. Seeds and oil, and stalks. Yields a yellow dye from the stems, which, "among dyes of an organic nature, rank next to the Persian Berry (*Rhamnus infectorius*), for the beauty and fastness of the dye."

Ord. FLACOURTIANEÆ. ARNOTTO FAMILY. (*Case* 5.)—An entirely exotic and mostly tropical Order, affording *Arnotto*, or *Annotto*, a red dye from the Bixa Orellana. Native of S. Am. *Fruit, seeds, cakes* from various countries, and *wood*. The fruit includes a thin pulp surrounding the seeds, which is collected and formed into *cakes*, or *flag*, or into *rolls*: the former comes chiefly from Cayenne, the latter from Brazil. In this country it is mostly employed in staining cheese and butter, for tinging varnishes, oils, spirits, &c., and for dyeing silk, and it is sometimes mixed with chocolate. The Caribs, and other Indians, paint their bodies with it. 4592 cwt. of Arnotto were imported to England in 1853. — *Mandingo Snuff-boxes*, the fruit of Oncoba spinosa? Gambia.—*Natal Snuff boxes*, apparently made of the same fruit.

Fig. 4.

CISTUS CRETICUS.

Ord. CISTACEÆ. CISTUS FAMILY. (*Case* 6.) — The Cistus Family is familiar to us from the many Rock-roses of our gardens, distinguished by the beauty yet short duration of their flowers. One of the most common is the *Gum-Cistus* (Cistus ladaniferus), so called because the whole plant, in hot weather, exudes a sweet, gummy, or glutinous substance, which has a strong balsamic scent.—*Ladanum* or *Labdanum*, a resin considered to be the product of Cistus Creticus (*Fig.* 4.). Crete and Syria. It possesses stimulant properties, but its use is obsolete with us. It is the *Ledon* of Dioscorides, in whose time the gum was obtained by driving goats among the shrubs, when the substance adhered to their fleeces and beards. Now that this gum is in greater demand, a peculiar instrument is employed to collect it, which is described and figured by Tournefort; and his accuracy is attested by Sieber, in his "Voyage to Crete." "It is a kind of rake

(*Fig.* 5.) with a double row of long leathern straps, employed in the heat of the day, when not a breath of wind is stirring. Seven or eight country-fellows, in their shirts and drawers, whip the plants with these straps, which, by rubbing against the leaves, lick off a sort of odoriferous glue sticking to the foliage." About 50 cwt. of it are annually collected in Crete, and sent to Constantinople. — *Gum* of Cochlospermum Gossypium, called *Gum Kuteera.* India. Properties similar to Gum Tragacanth, for which it is substituted in India. —*Leaf Bellows* made of the foliage of the Cochlospermum Gossypium, compactly stitched together, the handle and the snout alone being of Bamboo. They are used in smelting iron by the natives of the hill-country of the Soane river in India.—*Capsules and seeds of* Cochlospermum Orinocense. Brazil. Remarkable for the beautiful structure of the seeds.— Edible *Fruit* of Flacourtia cataphrasta, called *Lobi-Lobi Manis,* from Java.

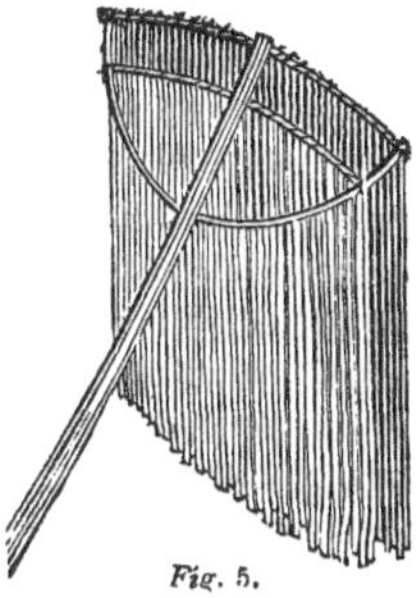

Fig. 5.
LADANUM RAKE.

Ord. Violaceæ. Violet Family. (*Case* 5.) — The *Violets* and the *Pansies* may be taken as the types of this Family: these have irregular petals; some tropical kinds, of whose properties we know little or nothing, have regular petals; but the former, whether of temperate or warm climates, are more or less employed medicinally, the roots possessing highly emetic properties. — *Cuicunchilli,* or *Cuichunchulli,* of S. Am. Roots of Ionidium parviflorum. Dr. Lindley determined that of Cuença to be from I. microphyllum. Emetic and purgative.—*Wine* of *Cuichunchulli;* a tincture of the root.—*Sweet Violet.* Viola odorata. Seeds and flowers. Eur. The roots are used medicinally, as emetic and purgative (as those of the *Dog Violet,* V. canina); but the plant is chiefly cultivated for the delicious odour of the flowers. Used also in the preparation of the officinal syrup; and as a test for acids and alkalies.

Ord. Moringaceæ. Horseradish-tree Family. (*Case* 5.) — This Natural Order, of doubtful position, is now generally placed near the Violet Family; but it requires a botanical eye to distinguish the affinities. It is confined to one genus, *Moringa.*—*Ben-oil, pods* and *seeds* of Moringa pterygosperma. An Indian tree, cultivated in Jamaica. Its pure fixed oil is much used by perfumers on account of its not easily becoming rancid, and by watchmakers, because it does not freeze. The roots have exactly the flavour of Horse-radish. Pods used in curries.—*Mocheris Gum.* Scinde. Exudes from wounds made in the bark of Moringa pterygosperma, and agrees in some of its properties with Gum Tragacanth.—*Ghalee*, seeds, so called, of Moringa aptera. Scinde. This species, if really distinct, is considered by De Candolle to yield the *Ben-oil*, to judge from his expression, "Been album offic." Probably a similar oil is afforded by both species.

Ord. POLYGALACEÆ. MILKWORT FAMILY. (*Case* 5.) —The common *Milkwort* of our heaths and dry pastures is a familiar example of this Family, of which the leaves, bark, and roots are, for the most part, bitter and astringent.—*Seneka-root*. Polygala Senega. N. Am. Diaphoretic, diuretic, and expectorant; in larger doses, emetic and purgative.—*Rhatany-root*. Krameria triandra. Native of Peru; and possesses powerfully astringent and tonic qualities. Said to be used for adulterating port-wine.—*Swan River Broom*. Comesperma scoparium. Swan River; where, as its botanical specific name indicates, it is used as a broom. From a small knotty root-stalk a quantity of slender twiggy branches arise, the whole forming a natural broom, which has only to be cut to be ready for use.

Ord. TAMARICACEÆ. TAMARISK FAMILY. (*Case* 5.)—This Natural Family is almost entirely confined to the genus *Tamarix*. S. of Eur. and Asia, possessing bitter and astringent bark.—*Manna of Mount Sinai* is an exudation from *Tamarix mannifera*, according to Ehrenberg, occasioned by an insect, a species of Coccus (*Coccus manniparus*) which inhabits the shrub: and this "Manna" consists wholly of pure mucilaginous sugar.— *Sakun*, or *Tamarisk Galls* of Scinde. Tamarix articulata. Highly astringent, used both in medicine and dyeing.—*Mahee*, from Bengal; apparently the same. Dr. Lindley says that such galls are produced by *Tamarix Indica, dioica, Furax* and *orientalis*; and that *Tamarix* (*Myricaria*) *Germanica* and *herbacea* are, occasionally, used medicinally.—*Wood* of Tamarix dioica. Handsome, close-grained.

Ord. CARYOPHYLLACEÆ. CHICKWEED FAMILY. (*Case* 5.)—The common *Chickweed* (Stellaria media), the *Pink*, the *Soapwort*, &c., are characteristic of this family, which, though numerous in genera and species, contains few plants possessing any marked properties. —*Clove July* (or *Gilly*) *flower*: the *flowers* of the *Clove Pink*, Dianthus Caryophyllus; used as a syrup.—*Soapwort*. Root and dried herb. Saponaria officinalis. Eur. Bruised and agitated in water, it raises a lather like soap, and may be used as a substitute for it.—*White Soapwort root*. Gypsophila Struthium. Eur.—*Arenaria rupifraga*. Tufts. Tibetan Himalaya. Interesting in connection with geographical botany, from growing at the greatest elevation of any phænogamic or flowering plant (in contradistinction to cryptogamic, such as Mosses, Lichens, &c.) in the world. Dr. Hooker's specimens were gathered at between 16,000 and 18,000 feet above the level of the sea: and the extraordinarily dense and tufted manner in which they grow (looking like, and by the unbotanical eye mistaken for, a compact Moss) indicates exposure in the clefts of rocks to extreme cold and driving hurricanes. One of the most elevated of flowering plants in our own mountains is nearly allied to this, and belongs to the same Natural Family, it is the Cherleria sedoides.

Ord. MALVACEÆ. MALLOW FAMILY. (*Cases* 6, 7, 8.)—An extensive Natural Family, chiefly inhabiting warm climates, eminently characterised by the mucilaginous properties, and the copious fibre of the inner bark. We have in England only the *Mallows*, *Marsh-Mallow*, and *Tree-Mallow*; but warm countries produce numerous species of *Hibiscus*, *Sida*, &c., which are generally shrubs and small trees, often with very handsome flowers; and in such regions alone, the *Cotton* is extensively cultivated, and forms a staple article of trade with almost all parts of the civilised world. It is the *fibre surrounding the seeds* which is so valuable, under the name of *Cotton*. None of the Malvaceæ are unwholesome; some are esculent; dye is extracted from others. The beautiful flowers of Hibiscus Rosa-Sinensis are used by the Chinese to blacken their eyebrows and their shoes.—*Marsh-Mallow.* Althæa officinalis. Dried flowers, foliage, &c. Europe. Demulcent and pectoral.—*Hollyhock.* Althæa rosea. S. Eur. Dried flowers. Mucilaginous and demulcent. The leaves dye blue.—*Ochro*, or *Ochra.* Hibiscus esculentus, *Pods* and *Fibre.* Tropical Am. and India. Cultivated for the sake of the green pods or seed-vessels, which are much employed in thickening soups, while the leaves are used for poultices. —*Cuba-bast* or *Lace-bark*, of Paritium elatum, as ascertained by Henry Christy, Esq. It is extensively used for tying cigars into bundles, and for tying up plants instead of Lime-bast.—*Rope* made of the fibre of Hibiscus tiliaceus; *Fibres* of Hibiscus Sabdariffa, and H. cannabinus, and cordage of Thespesia populnea, and from other *Malvaceæ*, are in the Collection.—*Musk-seeds.* Hibiscus Abelmoschus. These have a powerful musky odour, and are said to be roasted with coffee by the Arabs: tincture used against serpent-bites.—*Gingadan.* Sida sp. *Seed.* Used in Nubia and Soudan as a substitute for Coffee.—*Mallow.* Malva sylvestris. Dried herbage used as emollient and demulcent, and for poultices.—*Cotton*; from different supposed species of *Gossypium*, generally known as G. herbaceum, and G. Barbadense. Volumes have been written on the history, manufacture, and commercial importance of this plant; and the reader will find a valuable Treatise in the work of Dr. Royle. "On the Culture and Commerce of Cotton in India and elsewhere, London: 1851." It is, perhaps, not possible to give a more perfect notion of the magnitude of the Cotton trade to Great Britain, than by stating that, in the year 1853, besides 7,993,560 cwt. of raw cotton for our manufactories, there were imported, of cotton yarn, 1,126,741 lb.; of Cotton manufactures, Calicos and Muslins (white), 413 pieces; of dyed or coloured Cotton manufactures, 200,082 pieces; of Nankins, not dyed or coloured, 2194 pieces; of dyed or coloured Chiné, 23,998 pieces; of other Cotton articles, entered at value 436,132*l.*, besides Cotton Fringe, Gloves, Stockings and Socks, &c. &c., to a vast amount.* In regard to the fineness with which Cotton is now spun at Manchester, I would observe that J. Murray, Esq., of that city, in sending to our Museum three samples, in hanks, of different degrees of his very fine spun yarn, gives the following interesting information: "The comparative fineness of yarn is estimated by the number of hanks in a pound avoirdupois, and a hank is 840 yards in length.

A pound of No. 100. is	84,000	yards in length.	
" 300. "	252,000	"	
" 600. "	504,000	"	=286 miles.

"There is a very large quantity of No. 100. used for muslin, lace, &c. The No. 300. is used in small quantities, principally for lace, and must be considered a kind of luxury. The No. 600. is not available for dress, a small quantity was spun and manufactured into lace as a curiosity. Mr. Houldsworth spun a *very* small quantity for the Exhibition, of 2400 hanks in a pound; but it is too fine for anything but to be looked at. You will have to be very gentle in handling the No. 600., and even the No. 300. The cotton of which all the three specimens are made was grown in South Carolina." A large compartment on this floor is devoted to the material, with drawings of the plant: it contains—*Cotton-pods*, or ripe seed-vessels (generally still surrounded with the three-leaved involucre), more or less burst, and showing the cotton which envelopes the seeds; from various countries. *Nankin Cotton-pods* and *Cotton*, which is a variety naturally of that peculiar colour (not dyed, as is often supposed). *Raw Cotton.* From the Southern States of North America, Mexico, South America, East Indies, Africa, &c., as labelled.—*Cloth* of various kinds, and in various states of preparation, chiefly the work of natives in extra-European countries. A space is devoted to samples of *Egyptian, Macao, New Orleans, Surat* Cotton, from the bale; and cotton in various stages of manufacture, presented by Messrs. Glover and Dunn, of Manchester. Another part of the case contains *Cotton-pods* and *seeds*, together with *Oil* and *Oil-cake* from the waste seeds, prepared and presented by R. Burn, Esq., of Edinburgh. This oil-cake is strongly recommended in the "Quart. Journ. of Agriculture,"

* Board of Trade Report, 1853.

for feeding cattle. We are indebted to numerous other friends for contributing to this extensive Collection of Cottons, including *Maltese Cotton* and *Cloth*, *Dyak Cloth* of *Borneo*, *Cotton Cloths* of *Bhootan*, *Sikkim*; *Purses* from Soane River, &c. &c., and various articles from Dominica.

Ord. STERCULIACEÆ. STERCULIA FAMILY. (*Case* 8.)—In many respects nearly allied to the two preceding and to the two following Families: all possessing the same properties, viz., abounding in fibre and mucilage.—*Kola*, or *Cola Nuts*. Sterculia acuminata. West coast of Africa; cultivated also in the West Indies, and Bahia. The nuts or seeds have a pleasant, bitter taste and are much esteemed by the negroes as promoting digestion; they also prevent sleep, and are used by the native watchmen to keep themselves awake. Powdered *Kola* is applied to wounds or cuts. It was lately stated in a letter from Sierra Leone that 125 native traders had left that colony for the Gambia since the 1st of January, taking with them upwards of 1500 packages of *Kola-nuts*; a description of produce which is in great demand in the interior, and is now grown largely at Sierra Leone, although six years ago the trade was unknown. — *Kulheim*; fibrous bark of a Sterculia. Sikkim Himalaya. Used for cordage, &c. — *Oadal*; fibrous bark of Sterculia villosa. India. Of this are made fine pliable ropes, and especially those used by elephant-hunters in the jungles.—*Bag* of the Bark of Sterculia villosa. These bags are quickly made, by steeping logs, and stripping off the bark. Used for the conveyance of goods in the Goa teritory and Canara.—*Wood* of Sterculia platanifolia. Japan.—*Gum Tragacanth*, of *Sierra Leone*? Sterculia Tragacantha?—*Gum* of Sterculia Carthaginensis. Venezuela. —*Gum Kuteera*. Sterculia urens. Bombay. Resembling Tragacanth, and has been substituted for it (*Gum Kuteera* is by Dr. Lindley referred to Cochlospermum Gossypium, in *Cistaceæ*). — *Fruit* of Sterculia Balanghas. Malabar. Seeds wholesome, and when roasted nearly as palatable as chestnuts. In Amboyna the pericarp is burnt to make a pigment called *cassoumba*.—*Woven Cloth* from the fibre of Sterculia sp. Khasia. Worn by the Kuki chiefs. — *Bottle-tree* of Australia. Delabechia Australis. Interior of subtropical Australia.

Ord. BOMBACEÆ. SILK COTTON-TREE FAMILY. (*Cases* 8, 9.)—Nearly allied to the last, and possessing the same properties. It includes large trees of tropical climates of the Old and New World, which are adorned with very large handsome flowers. Trees, generally speaking, bear insignificant flowers: those of this Family are remarkable for their great size. — *Baobab*, or *Monkey Bread-fruit*. Adansonia digitata. (*Fig.* 6.) Fruits and flowers, in liquid. Trop. Africa. Lake Ngami. The product of one of the most remarkable trees in the world, of which, and its flowers, there are drawings on the staircase. Trunks have been measured by Adanson on the coast of Senegal, thirty feet in diameter; the height rarely exceeds eighty feet. The wood is pale, light-coloured, and so soft, that in Abyssinia the wild bees perforate and lodge their honey in the trunk, which honey is considered the best in the country. On the west coast the trunks are hollowed by the natives, and the dead deposited therein, where they become mummies. The fruit is perhaps the most useful part of the tree: its pulp is acid and agreeable; and the juice, mixed with sugar, constitutes a drink that is deemed a specific in putrid and pestilential fevers, and is therefore an article of commerce.—*Fruit* and *wood* of the *Gouty-stem tree.* Adansonia of N. Australia, (*A. Gregorii.*)—*Flower* of Pachira alba, in liquid. Braz.—*Silk* or *Down* and *Fruit* of Bombax Munguba. Braz. Down used for stuffing cushions.—*Silk* and *Fruit* of Eriodendron Samauma. Braz. This Silk Cotton is used by the Indians of the Amazon for wrapping round the ends of the arrows used with the blowpipe, and for stuffing cushions.—*Silk Cotton* from Bombax Ceiba. B. Guiana. Exported to the States, and used in the manufacture of hats.—*Down* and *Flower-buds* of *West-India-Cotton*, from Eriodendron anfractuosum. S. Am. The large stems of this tree are hollowed out to form canoes. The wood is soft, and subject to the attack of insects; but if steeped in strong lime-water, it will last for several years, even when made into boards and shingles, and in situations exposed to the influence of the weather. The cotton is employed in stuffing mattrasses, and is said to serve as well as feathers. The caterpillar of the Macaca beetle, considered by some, when gutted and fried, as a very great delicacy, abounds in the decayed stems of this tree.—*Cloth* woven near Gowhatty, Assam, from the cotton of the *Simool*, Bombax heptaphyllum.—*Silk Cotton*, and the pods which contain it, of Ochroma Lagopus. Jamaica. Used in the manufacture of hats; and *Nest* of the "Doctor Humming Bird," made of the same. *Wood* of ditto, called *Cork-wood* from its lightness; and inner bark of ditto.—*Durion* or *Durian Fruit.* Durio Zibethinus. Malay Islands. Is considered one of the most delicious productions of nature; but fœtid, and therefore disagreeable to those unaccustomed to it.—*Durion* or *Durian Fruits* of *Ceylon*. Cullenia excelsa. Fruit not eatable, but monkeys are very fond of it.—*Capsule* of Durio Oxleyana. India.—*Capsules* of Durio Ootan. Malacca.—*Fruit* of Matisia cordata. New Grenada. Fruit is edible, with a flavour like

Fig. 6.

ADANSONIA DIGITATA.

Apricots.—*Hand-plant*, flowers in liquid. Cheirostemon platanoides. A sacred plant among the ancient Mexicans; considered so probably in consequence of the resemblance of the stigma of the flower to the human hand. — *Capsules* of Neesia altissima. — *Fibrous bark* of Hoheria populnea. N. Zeal. The bark, like that of Mallows, affords a demulcent drink used in medicine, and a cordage.—*New Zealand Cotton*, fibre of Plagianthus urticinus.—*Currajong bark*, and netting made of it. Plagianthus sidoides. Moreton Bay.

Ord. Byttneriaceæ. Chocolate Family. (*Case* 9.) — By Jussieu this Order is united with *Sterculiaceæ* and *Bombaceæ;* it is peculiarly interesting, as including the Chocolate-tree.—*Chocolate*. Theobroma Cacao. *Fruits* and *seeds* (the latter yield Chocolate and Cocoa) from Trinidad, Granada, Bahia, Nicaragua, Caracas, Pará, Guayaquil; shells of the "Nuts," *Nibs;* numerous samples of prepared *Chocolate*, and *Cocoa;* a very fine collection, prepared and presented by Messrs. J. and S. Fry and Sons, of Bristol; another by Mr. L. Monteiro, and another by Mr. G. B. White. Fruits that have ripened in England (in the stove) are presented by His Grace the Duke of Northumberland (from Syon), and by Dillwyn Llewelyn, Esq. A large drawing on the staircase (besides the living plant in the Garden) will give an idea of the foliage, flowers, and recent fruit. The fruit is large, and contains several seeds, with a crustaceous shell or husk (*testa* or *integument* of the seed), imbedded in pulp. From the entire seed, including the husk, *Cacao* (or *Cocoa*, as it is generally called) is prepared:—from the seed, after the husk is removed, *chocolate*. The latter is made into a paste, and mixed with *Vanilla*, &c., and generally coloured with *Arnotto*, and dried in cakes in the form in which we see it in the shops. The husks or shells of the seed are largely employed as a substitute for the more expensive beverage. In 1853, 8,283,694 lb. of Cocoa, 29,082 lb. of Cocoa paste or Chocolate, were imported into England.— *Goorakhee Khorai Fibre*, Rope, and Yarn, from Abroma augusta. Assam.—*Fruit* of Herrania pulcherrima. Peru. Of botanical interest only.— *Guasino colorado bark;* of Guazuma tomentosa, and *Guasino blanco*, probably a Guazuma. Venezuela.

Ord. Tiliaceæ. Linden or Lime Family. (*Case* 9.) — A Natural Family again much allied to the four preceding groups, and possessing, like them, mucilaginous properties and abundant fibre in the inner bark, yielding many useful products. *Corchorus olitorius*, as its name implies, is much cultivated in Egypt and elsewhere as a potherb. — *Lime Flowers*. Tilia Europæa. Eur. Antispasmodic. *Bast*, or *Russian Matting*, and *English bast*. This is the inner bark of the Tilia Europæa, generally prepared in Russia.— *Paper* made of inner bark and rope of *Bast-wood*. Tilia Americana. N. Am.—*Jute*

Fibre and *Dowlas*, *Carpets* and *Gunny Cloth* (for making ricebags in India), made from Corchorus capsularis, and probably other species. The fibre is extensively used in England: in 1853, 275,578 cwt. were imported.—*Jute Paper.* Five samples of as many excellent kinds of Papers, recently (1854) prepared at Madras, under the direction of Dr. Hunter; made from old Gunny bags.—*Sabots de Filasse;* made in the S. of France of Jute and Cotton.—*Gootee Fibre*, *Rope*, and *Yarn*. Triumfetta angulata. Assam.— *Bur-weed* fibre. Triumfetta semitriloba. Jamaica.— *Inner Bark*, and *Rope*, made of Apeiba Petouma. S. Am.—*Fruit* of Apeiba aspera. Trop. Am.— *Fruits* of Sloanea Jamaicencis.—*Rosaries* used by Indian devotees, made of the seeds of Elæocarpus Ganitrus. Sold in London shops for necklaces.— *Buah Kitjapie.* Fruit of Elæocarpus glaber. Java.

Ord. Dipterocarpeæ. Malay-Camphor Family. (*Case* 9.)—Stately forest-trees of the Indian Islands, yielding a resin (generally denominated *Dammar*), of which the most remarkable is the *Borneo* or *Sumatran Camphor*. In its natural state it is deposited in crystals, in cavities of the trunks of Dryooalanops Camphora.—This costly substance was spoken of by the celebrated traveller, Marco Polo, before the year 1299, as "*Canfara Fansuri*" of Sumatra; and the poet Camoens, 272 years later (1571), sings—

> "Bornëo here expands her ample breast,
> By nature's hands in robes of *camphire* drest;
> The precious liquid, weeping from the trees,
> Glows warm with health, the balsam of disease."

Its nature is very different from true camphor; it sells for 35 dollars the catty, and is chiefly used for embalming the dead. A section of a hollow trunk exhibits *in situ* the crystallised camphor, called *Kassur-Barus.*—*Mimak-Kassur*, or *Camphor-oil*, of the same.—*Leaves* and *unripe fruit* of Dryobalanops Camphora, in liquid; and drawing.—*White resin* of the same.—*Fossil wood*, considered to be of the same tree, almost partaking of the nature of coal, containing the crystals; all the above collected and presented by J. Motley, Esq., Borneo.—*Gum Piney*, a resin from Vateria Indica, and *Piney Tallow*. Malabar and Ceylon. From this, Piney varnish is made, which resists the action of water; and *beads*, resembling amber beads; it is electric when rubbed.— *Resin* and *bark* of Dipterocarpus turbinatus. E. Bengal. It yields the *wood-oil*.—*Hora dommeli*; a *Resin*, and *Wood-oil*, the former mixed with Paddyhusks for burning: both from Dipterocarpus lævis. Canara, and Ceylon.—*Indian Balsam of Copaiba*, essential oil.—*Doon*, or *Doongaba* (Doon-tree) *resin*, of Doona Zeylanica; used as incense, makes a good varnish, and the wood is excellent. Ceylon.— *Resin* of Doona Gardneri. Ceylon.

Ord. TERNSTRŒMIACEÆ. TEA FAMILY. (*Cases* 7, 8.)—As including the Tea-plant, the present Natural Order may unquestionably be considered an important one, affording to millions the drink "that cheers but not inebriates." We have no space to enter on its history, botanical and commercial. Three kinds of Tea-plant are in extensive cultivation: viz. *Green* (Thea viridis) *Black* (T. Bohea), and *Assam* (T. Assamica). Mr. Fortune has stated that *black* and *green tea* of the shops may be made from either or both of the first two species. The importation of Tea in G. Britain was, in 1853, 70,735,135 lb. The Museum collection is extensive and valuable.—Samples of *various sorts* of the ordinary, but good teas of common use, from Messrs. Twining: these are, of BLACK: *Congou* (three qualities), *Souchong, Plain Caper, Chulan Caper, Orange Pekoe, Scented Pekoe, Flowery Pekoe, Assam Souchong, Assam Pekoe*; of GREEN: *Twankay, Hyson* (two qualities), *Young Hyson* or *Pontazan* (two qualities), *Imperial*, and *Gunpowder Hyson.*—*Large* and *Small Black Ball tea:* i. e. Tea made into balls; those of a small size wrapped in Bamboo paper; the larger ones enveloped in the flowering sheaths of Indian Corn, to preserve the aroma.—*Old Man's Eyebrow Tea.* This is done up in short twisted sticks, and perhaps bears allusion to the legend of some Chinese saint tearing off his eyebrows and throwing them on the ground, where they sprouted into tea-plants; and representations of this wonderful transformation are seen on Chinese screens, &c.—*Brick Tea,* made of old damaged leaves and stalks pressed and dried in moulds, much used throughout Tibet and in Asiatic Russia, boiled with salt and oil, and thickened with flour.—*Native Tea-pots* from Shigatzi, used by the Tibetans.—*Wheatsheaf Tea,* made in sticks or long pieces and tied up in small bundles.—*Tea* made expressly for the *Emperor of China,* presumed to be super-excellent.—*Assam Tea,* from the Assam Company (six samples), and six from Chinese Tea-plants grown in Assam; viz. *Hyson,* three qualities; *Congou,* first class; *Souchong,* first class; and *Flowery Pekoe.*—*Kamoun Tea,* green and black. This and Assam Tea are made in the East India Company's possessions, and command an extensive and increasing sale.—"*Medicated Tea,*" from Chinese Tartary, probably the *Pu-'rh* tea, mentioned by De Guignes, prepared, according to Fortune, from various herbs, and used for medical purposes.—*Teas* from *Chittagong,* made from Assam and Chinese plants.—"*Extract of Tea*": a preparation made by the Chinese in lozenges of various sizes and stamped with various devices, and used to dissolve in the mouth on long journeys.—*Paper Tea.*—*Yellow Tea,* from the great Russian Fair of Nijni Novogorod.—*Atkalah Tea,* and *Tea* from *Yarkand,* probably Chinese teas made for those markets.—*Chinese Tea-cup,* of elegant make. — *Tea-seeds;* and *Tea-seed oil:* the latter from Java. — *Theine*, the principle of Tea (which exists also in Coffee and other plants). — *Flowers of Tea.* — A case of the various ingredients employed in the manufacture of *Green Tea* by the Chinese: viz. Turmeric, Gypsum, and Prussian blue. — *Samples* of various *spurious Teas*, manufactured in England, and also of adulterated Chinese Teas, named (both from A. Hassall, Esq.)—The Museum contains an elaborate series of Chinese drawings, on an adjacent wall, explaining the origin and cultivation and manufacturing of Tea, of which the first drawing represents a monkey upon a rock gathering the Tea-plant, and showering down specimens on the gaping Chinese below.—Closely allied to the Teas are the Camellias. Of these we have *flowering specimens* of Camellia Japonica, modelled in wax. — *Fruits* of Camellia Japonica. — *Tea-oil*, extracted from the *seeds* of Camellia Sasanqua. — *Bark* and *wood* of *Caraipé* or *Pottery-tree*, Caraipa augustifolia; *Burnt bark* of the same, prepared to make the ware, and *specimens* of *the Pottery*. Pará. This was first brought into notice by Aublet: "Les Gueripons employent les cendres de son écorce mêlées avec une terre grasse, pour leur poterie." They will bear almost any amount of heat. (Of late, Mr. Spruce has had reason to believe that the Pottery-tree of Pará belongs to the Nat. Ord. *Chrysobalaneæ.*)

Ord. Olacineæ. Olax Family. (*Case* 9.) — *Wood* of Ximenia Americana. Tropical Amer. and India. Substitute for Sandal-wood. Flowers very fragrant, smelling like cloves.

Ord. Aurantiaceæ. Orange Family. (*Cases* 9, 10.)—All are familiar with the types of this family, Oranges and Lemons, which, even in a commercial point of view, are of no small importance; seeing that in 1854, these fruits, to the extent of 814,065 bushels, computed at 367,074*l.* worth in value, were imported into England. They are looked upon as the golden fruits of the Hesperides, whence Jussieu called this family Hesperideæ. Handsome ever-green trees of India, with fragrant flowers, leaves compound or jointed on the petiole, and the whole plant abounding in glands filled with fragrant essential oil. — *Wampee fruits.* Cookia punctata. China and Molucca. Fruit, with the flavour of the grape.—*Lime berries*, of Triphasia aurantiola. Manilla. —*Elephant*, or *Wood-apple*. Fruit of Feronia elephantum. India. Edible. The tree yields a gum much resembling gum Arabic. — *Bael fruit*, Indian *Bael*, *Bêl* or *Bêla*, *Bengal Quince;* sliced and dried unripe fruit of Ægle Marmelos. India. Ripe fruit delicious, but laxative, when unripe, a remedy against dysentery and diarrhœa, and a popular medicine in England. *Diwool gum.* Ægle Marmelos. Ceylon. *Bark* of the same, used on the Malabar coast medicinally. — *Citron*, fruit of Citrus medica. Rind very thick, much used for preserves; yields a considerable proportion of essential oil.— *Oils* of *Citron* and *Cedrate.*— *Fingered Citron;* a remarkable variety of C. medica, reminding one of the " Fingers and Toes" in the Turnep. China. —*Forbidden Fruit.* Citrus Paradisi. Jamaica. —*Lemons.* Fruit of Citrus Limonum. Invaluable for preventing scurvy in long voyages, used also as a mordant in fixing vegetable reds. There are numerous varieties.— *Oil* of *Lemon-peel.*— *Limes;* fruit of Citrus Limetta. Of this species or variety there are several kinds. The acid is more abundant and purer than in the Lemon, and is celebrated in the preparation of Glasgow punch.—*Madagascar Lime,* var. received from the Mauritius. — *Comquat,* a most curious small nutmeg-shaped orange, Citrus olivæformis. China.—*Bitter* or *Seville Orange.* Citrus vulgaris. By some considered the wild stock of the eatable orange, which latter is supposed to owe its sweet flavour to cultivation, and to hold the same relation to it that the apple does to the crab. Excellent for marmalades, which is the crushed fruit, boiled with sugar. From the flowers, *Neroli* and orange-flower water are obtained.—*Sweet Oranges:* fruit of Citrus Aurantium. Among the numerous varieties are the *St. Michael's,* the *Blood-red,* the *Maltese,* the *Mandarin,* and the *Majorca,* or seedless state. — *Orange-berries,* or immature fruits, used as "issue peas."— *Orange* flowers and leaves, dried.— *Orange-wood,* and *Toothpicks* made of the same from Madeira and Rio Janeiro, and *Walking-cane* of the same.— *Oil,* prepared from Orange-seeds, Jamaica. — *Otto of Neroli. Oil of Sweet Oranges. Orange-flower-preserves* from Bagdad. *Navel Oranges.* — *Bergamot Orange,* or *Mellarosa.* Fruit of Citrus Bergamia; somewhat pyriform; rind extremely fragrant. and often made into boxes. Flowers and fruit yield the essential oil of Bergamot, much employed by perfumers.—*Shaddock.* Citrus decumana; the largest of all the orange tribe, called Pompelmousse by the French. *Oil* of ditto. Many of the above orange-group are successfully cultivated in Devonshire, with no fire heat, and with very little winter covering, particularly by J. Luscombe, Esq., of Coombe Royal, to whom we are indebted for fruits so reared.

Ord. HYPERICINEÆ. TUTSAN FAMILY. (*Case* 10.)—Two kinds are employed medicinally, stem and foliage, viz. *Common St. John's Wort,* Hypericum perforatum, and *Common Tutsan,* or All-heal, H. Androsæmum, both European.

Ord. GUTTIFERÆ. GAMBOGE FAMILY. (*Case* 10.)—Called *Guttiferæ,* from the resinous, acrid, and purgative juices which flow from them. Some bear esculent and delicious fruits.—*Fruit* of Clusia flava. S. Am. Martius considers that this tree yields the Hog-gum; but certainly not the *Jamaica* Hog-gum.—*Resin* of *Ducu,* Clusia sp. Ecuador, where it is used for incense in churches.

—*Mangosteen*. Fruit of Garcinia Mangostana, dry and in liquid. Moluccas. These fruits only come to perfection in the Malay Archipelago, and are reckoned by travellers the most delicious in the world.*—*Kokum butter*, from Garcinia purpurea. Scinde. —*Gamboge butter*. Garcinia pictoria. Mysore.— *Pipe* or *Siam Gamboge*, considered the best of the kind ; a highly drastic purgative, and valuable colour for painters; supposed to be derived from Garcinia Cochinchinensis. — *Lump Gamboge*. Ceylon. Said to be from *Gambogia Gutta*. We have yet much to learn relative to the trees yielding gamboge.—*Fruit* of Xanthochymus pictorius. India. Yields a large quantity of indifferent gamboge.— *Naghesur*, or *Nagassar*. Flower-buds of Mæsua ferrea. India. Used in medicine and for scenting clothes.—*Varnish-oil* of the same. Sylhet. The wood is said to be very hard.—*Suringee*, or *Gordeondys* ; the flowers of Calysaccion longifolium, which are collected and dried in Bombay, and exported to Bengal for dying silk. — *Gum* of Calophyllum inophyllum. Ceylon. Both resin and an oil called *Poon-seed oil*, are yielded by the seeds of this plant, and the latter is supposed to be the true *Tacamahaca* of Bourbon.—*Mammee Apple;* the fleshy esculent fruit of Mammea Americana, and the *gum*, which is used for destroying the Chiggers (*Culex penetrans*) in the negroes' feet. W. Indies.—*Anani*, *Oanani*, or *Mani gum ;* a resin derived from Moronobea globulifera. Pará. A plant closely allied to, if not identical with, the following.— *Hog-gum*, from Moronobea coccinea. S. Amer. and Jamaica (Dr. M'Fadyen) ; a resin resembling Burgundy pitch, and used as a substitute for it in the W. Indies, and for making torches. —*Peraman*. Resin of a Moronobea of the Orinoco.— *Butter* or *Tallow Tree*. Fruit. Pentadesma butyracea. Sierra Leone. Yields a yellow fatty substance.

Ord. ERYTHROXYLACEÆ. COCA FAMILY. (*Case* 11.)—The name of this order is derived from the red colour of the wood of some species. One alone is particularly worthy of notice.—*Coca*. *Leaves*, loose and in bundles, of Erythroxylon Coca. Peru and Columbia.—*Powdered leaves* of the same from the Rio Negro, and the *Popona*, a small gourd used in holding lime for chewing with Coca. Coca is most extensively used by the labouring classes, especially the miners of Peru, for its remarkable powers in stimulating the nervous system ; in that respect resembling opium. We have published a figure and description of it in the "Companion to the Bot. Magazine," vol. ii. p. 25. tab. xxi. The use of it may be traced as a masticatory to the

* In 1855, this fruit, for the first time in Europe, ripened at Syon, His Grace the Duke of Northumberland's. We were privileged to partake of it, and can testify that it deserves the high encomiums travellers have pronounced upon it. One of them was presented to this Museum.

times of the earliest Incas, when it was only employed in religious ceremonies. Now it is universal, and when chewed with a small mixture of ashes of Quinoa or Cactus, or lime, it is said to relieve hunger in a remarkable degree, but when taken in excess, to produce, according to Dr. Pöppig, worse effects than those occasioned by opium-eating.

Ord. MALPIGHIACEÆ. BARBADOES-CHERRY FAMILY. (*Case* 11.)—The fruits of some species are eaten. Hairs, when present, are often stinging and curiously fixed by the centre.—*Duka Bark;* Byrsonima? sp. B. Guiana. Used probably for tanning, as that of Murici is.—*Caápi.* Stem of a plant of the Natural Order from Rio Uapés, Brazil, used in preparing an intoxicating beverage.— *Murixi bark*, of Byrsonima spicata. Pará.

Ord. HIPPOCRATACEÆ. HIPPOCRATEA FAMILY. (*Case* 11.)—*Kokuna resin, seeds* and *bark;* from Kokoona zeylanica. Ceylon. The yellow bark used as cephalic snuff, and sold in the Bazaars.

Ord. SAPINDACEÆ. SOAP-BERRY FAMILY. (*Case* 11.)—Trees and shrubs, principally of S. Am. and India, yielding edible fruits; of some the leaves are poisonous.—*Soap-berries*. Fruit of *Sapindus saponaria*, used in making a lather, instead of soap, and for making Rosaries. India. — *Soap-nut* oil, of Sapindus emarginatus. Tanjore.— *Buah Rarak*, fruit of Sapindus Rarak; and *Buah tampyang*, fruit of S. rubiginosus. Java.— *Litchi*, esculent fruit of Nephelium Litchi. China. Much eaten by the Chinese and by residents in India generally. Not unfrequently imported, and sold in fruit-shops in London.—*Longans;* esculent fruits of Nephelium Longan. China and Malay Islands. Edible, but inferior to the Litchis. — *Ramboostans*; esculent fruits of Nephelium lappaceum. Malay islands. Much eaten in Java and Sumatra.—*Akee;* remarkable and esculent fruit from tropical Africa, of Blighia sapida: *model* of ditto and *oil* ditto. —*Fruit* and *oil* from the seeds of Alectryon excelsum. N. Zeal. — *Snake Nuts* from British Guiana, of Ophiocaryum paradoxum. The embryo resembles a coiled snake, and has thus obtained a certain degree of celebrity. — *Horse-chestnuts.* Seeds of Æsculus Hippocastanum. Persia. They contain a great deal of starch.— *Wood* of Koelreuteria paniculata. China. — *Guarana;* Paullinia sorbilis. Brazil. From the seeds of this plant, guarana is prepared by the natives and used as food and medicine.—A *tongue* of a *fish* called "Pirarucu," employed as a rasp for grating the Guarana. Amazon. *Supple-jack sticks;* Paullinia Curassavica. W. Indies.

Ord. ACERACEÆ. MAPLE FAMILY. (*Cases* 12, 13.)—Our common Sycamore, or Great Maple (Plane Tree of the Scotch), is characteristic of this family; the wood of many kinds is valuable.—*Maple Sugar* (in cakes), derived from Acer saccharinum. (*Fig.* 7. U. S. and Canada.) This sugar is universally prepared by boiling down the juice that is tapped in the spring from this Maple. It is seldom refined; but in lump or cake it is much preferred for many purposes to cane sugar. — *Woods* of *Common Maple,* Acer campestre, and a *butter stamp,* Eur.; of *Red Maple,* Acer rubrum, and a variety called *Curled Maple,* N. Am.; of Acer Pseudo-platanus, Eur.; of Acer dasycarpum, &c. — *American Clothes-peg,* made of Maple wood. — *Paper* made of *Maple wood.* U. St. Am.

Fig. 7.

ACER SACCHARINUM.

Ord. Rhizobolaceæ. Souari-nut Family. (*Case* 13.) —*Flowers* and *eatable Fruits* of Caryocar nuciferum, S. Am. Often sold in London shops. Kernel large and well-flavoured; sometimes called butter-nut.

Ord. Meliaceæ. Melia Family. (*Cases* 13 & 14.) —Somewhat allied to Sapindaceæ; bitter, astringent, and tonic.—*Gum* and *Wood, Seeds,* and *necklace,* and oil called *Margosa Oil,* from Azadirachta Indica, India.—*Necklace* of the seeds of Melia Azedarach, *L.*, or "Pride of India." A tree much cultivated in warm countries, for its beauty and the shade it affords.— *Soontool. Fruit* of Sandoricum Indicum. Malay Islands. Called "False Mangosteen" by the French. The pulp has a pleasant acidity; but is improved by being made into syrups, jellies, and preserves. — *Doocoo,* fruit of Lansium domesticum. One of the many esculent fruits of Java. — *Crab-oil* and *wood* of Carapa Guianensis. S. Am. The oil is obtained from the seeds; it is very bitter, and burns well. The Indians smear their persons with it. Its excessive bitterness repels insects. — *Coondi nuts,* of Carapa Guineensis, probably not different from the American Carapa. —*Wooden lock* made of the wood of the same by the negroes.

Ord. Canellaceæ. Canella Bark Family. (*Case* 13.)—*Canella bark,* Canella alba, by many considered to form a distinct order (Canellaceæ). Jamaica. A stimulant tonic, and used as a spice. Formerly confounded with Winter's bark, Drimys Winteri.

Ord. Humiriaceæ. Umiri Family. (*Case* 13.)— *Umiri bark,* Humirium floribundum. Yields a fragrant balsam. Pará. *Mimiac chulan;* an oil from the flowers of Aglaja odorata. Amboyna.

Ord. Cedrelaceæ. Mahogany Family. (*Cases* 14, 15.) —Trees with alternate pinnated leaves, of tropical Ind. and Am. Bark of some used in intermittents, as Soymida febrifuga, and Swietenia febrifuga.—*Mahogany,* wood and seed-vessels. Swietenia Mahogani. One of the most valuable of all woods for furniture. A single trunk or log has been sold for 292*l.* 12*s.* 27,495 tons were imported in 1853, chiefly from Honduras. See a full account of this tree in Hook. Bot. Misc., i. p. 21., tab. xvi. xvii.—Model of a truck, with a small block of mahogany, and of yokes, &c. for the oxen employed in bringing the mahogany logs to the coast; presented by Chief-Justice Temple of Honduras.—*Satin-wood.* Wood of Chloroxylon Swietenia. Remarkable for its colour and veining.—*Wood* of Cedrela odorata, a valuable timber; and its *gum* called W. Indian Cedar-tree gum.—*Cedar-wood* of *N. S. Wales.* Cedrela Toona. Found also in India.—*Fruits* and *wood* of Oxleya xanthoxylon, and of wool dyed with the wood; and of Flindersia australis. Remarkable in structure; both from Australia.

Ord. AMPELIDEÆ. VINE FAMILY. (*Case* 15.) —Important though the Vines, the type of this small family, are, in a commercial view, in the shape of wine, brandy, &c., there is little that can be instructive in a museum.—*Argol*, or *impure Cream of Tartar* (or *acid tartrate of Potash*), as it is taken in the form of an incrustation from the interior of wine casks; used in dyeing, in medicine, &c.; and to such an extent that 52,757 cwt. were imported from wine countries for home consumption, in 1853.—*Currants of Zante*, or *Levant Currants;* which are not, as many suppose, analogous to our garden currants, but are a small stoneless grape, Vitis vinifera, (see drawing on the staircase,) originally from Corinth, dried. 267,265 cwt. were imported in 1853.—*Raisins, Muscatel*, from Spain; *Cape, Valencia, Malaga, Sultana* (without stones) from Turkey: these are dried varieties of the more perfect fruit of the same vine, V. vinifera. Of fresh grapes 15,154 bushels were imported in 1853, besides 251*l.* worth entered at value, and of Raisins 446,997 cwt.

Ord. GERANIACEÆ. CRANESBILL FAMILY. (*Case* 15.)—Every one is familiar with the Geraniums and Pelargoniums, the latter especially in favour with cultivators, and very numerous in S. Africa. They are astringent and slightly aromatic and resinous, and used in popular practice. So resinous is one species, Geranium spinosum, that the dead stems become masses of resin in the sands of S. Africa, retaining their form, and they burn like a torch, giving out a most agreeable odour.—*Cranesbill-root*, and *Extract.* Geranium maculatum. N. Am., where it is much employed, and said to contain more Tannin than Kino does.—*Herb Robert.* Geranium Robertianum. Dried herb, used in rustic practice. Eur.—*Oil* or *Attar* of Geranium, used for adulterating Attar of Roses; said to be from the Rose-leaved Geranium; Pelargonium odoratissimum.

Ord. OXALIDACEÆ. WOOD-SORREL FAMILY. (*Case* 15.)—Remarkable for the agreeably acid properties, due to the presence of oxalate of Potass. — *Fruits* of *Blimbing,* Averrhoa Bilimbi; and *Carambola,* Averrhoa Carambola. India. Esculent, the former eaten raw; the latter, as olives. — *Wood-sorrel.* Oxalis Acetosella. Eur. Considered to be the genuine *Shamrock;* dried herb a refrigerant, and fresh as a salad, antiscorbutic. *Oxalate* of *potash* from the same.— *Oca root,* of Chili. Oxalis crenata. One of the many roots recommended as a substitute for potato.

Ord. Linaceæ. Flax Family. (*Cases* 16, 17.)—Linum, or *Flax,* is almost the only genus in this family one, and a most important one it is, as exhibited in the *Common Flax,* L. usitatissimum. Eur. Its products occupy two cases, nearly all being the growth and manufacture of Ireland, presented by the Royal Irish Flax Society, and showing the various processes which the raw material undergoes, till it is perfected in the beautiful *Linens,* &c., for which Ireland is so celebrated. There are besides, the *Oils, Oil-cake, Linseed Meal,* &c., the products of the seed; and *patent* Oil-cake made of Linseed marc, and the refuse of Flax after the fibre has been extracted. Notwithstanding the continually increasing growth of the plant in our own country, there were from the Continent imported into England in 1853, of dressed Flax 26,726 cwt., of undressed ditto 1,556,720 cwt., and of Tow and Codilla of Flax 299,928 cwt. exclusive of the enormous amount of manufactured linens; and 1,035,335 quarters of seed! A substance resembling Cotton is now prepared from refuse Flax. — *Purging Flax.* Linum catharticum. Dried plant used in rustic practice. Eur.

Fig. 8.

GUAIACUM OFFICINALE.

Ord. Zygophyllaceæ. Guaiacum Family. (*Case* 18.) —Mostly warm country plants, possessing a stimulant resin, some used medicinally. —*Guaiacum bark, chips, Gum Guaiacum,* and *Lignum Vitæ;* the products of Guaiacum officinale. (*Fig.* 8.) W. Ind. The wood much prized for its hardness.—*Gum Guaiacum tears,* from Guaiacum sanctum. Mexico.

Ord. Rutaceæ. Rue Family. (*Case* 18.) — Usually provided with resinous dots or reservoirs, abounding in a bitter odoriferous essential oil. Antispasmodic, febrifuges, and tonics. —*Bucku leaves;* from Barosma serratifolia, B. crenulata, and B. betulina. Cape. — *Rue; herb* and *seed.* Ruta graveolens. Eur. — *Angostura Bark.* Galipea Cusparia. B. Guiana. — *White Dittany root.* Dictamnus albus. S. of Eur.

Ord. Xanthoxylaceæ. Xanthoxylon Family. (*Case* 19.) — Diaphoretic, febrifuges, and tonics, owing to an aromatic and pungent volatile oil. — *Prickly-Ash bark.* Xanthoxylum fraxineum. U. S. —

Stems of Xanthoxylon Clara Herculis, and *walking-sticks* of the same from China.—*Japan Pepper*. Fagara piperita.—*Resin* of Ailantus Malabarica, &c. E. Ind.

Ord. SIMARUBACEÆ. QUASSIA FAMILY. (*Case* 19.)—Trees and shrubs, with an intensely bitter principle; much employed in consequence medicinally and otherwise. — *Surinam Quassia wood*. Quassia amara. S. Am. Now, it is said, seldom imported.—*Quassia wood* of the English Mat. Med. Picræna excelsa. W. Ind.; often called *Bitter Ash* and *Bitter wood*.—*Drinking-cups* of wood of the same, used to give a bitter flavour to the liquid.—*Simaruba bark* is from Simaruba amara. B. Guiana. — *Fruits* of Brucea antidysenterica. Abyssinia. — *Cedron*. The large cotyledons of Simaba Cedron. N. Grenada, where it is considered to supersede the sulphate of quinine.

Ord. CELASTRACEÆ. SPINDLE-WOOD FAMILY. (*Case* 20.) — The common Spindle-wood of our hedges may be considered characteristic of this Family. Properties subacrid. Celastrus venenatus is reckoned poisonous. — *Wood* of the *Spindle-tree*, Euonymus Europæus, for making skewers and spindles. Yields excellent charcoal for drawing. — *Malkamque seeds*. Celastrus paniculatus. India. Used medicinally.

Ord. RHAMNACEÆ. BUCKTHORN FAMILY. (*Case* 20.) —Cathartic and emetic. — *Persian* or *Yellow Berries*. Rhamnus infectorius, S. Eur. Much used in calico-printing, yielding rhamnine and rhamneine. 7192 cwt. were imported in 1853. — *Buckthorn; seed, berries, juice,* and *wood* of Rhamnus catharticus. S. Eur. — *Sap-green* is a well-known paint, made of the colouring matter of the berries of R. catharticus mixed with lime. — *Wood* and *Jujubes* of Zizyphus vulgaris: the fruit affords the pulp of which the lozenges called Jujubes are made. Z. Lotus is one of the Lotuses of the ancients. Z. Spina Christi is considered by some to be the thorn with which our Saviour was crowned.—*Chaw-stick* and *powder* of Gouania Domingensis, used for cleaning the teeth: the powder is a dentifrice.

Ord. ANACARDIACEÆ. CASHEW-NUT FAMILY. (*Case* 21.) — An important family, and, owing to the presence of an acrid resinous juice, often poisonous. Several yield esculent fruits.—*Pistachio nuts*. Pistacia vera. S. Eur. Much sold in fruit-shops. — *Gum mastic;* a concrete resinous juice of Pistacia Lentiscus. — *Chinese* and *Bokhara galls;* produced on leaves of Pistacia Lentiscus? — *Chian Turpentine*. A liquid resinous exudation from Pistacia Terebinthus. Chio or Scio, and N. of Africa. Diuretic and excitant. — *Cashew-nuts*. The well-known fruit of Anacardium occidentale, with its curious pear-shaped peduncle, very acrid; the kernel eatable when roasted; also *Gum* of the same, or *Cadjii Gum* and *Oil*. S. Am.

—*Nuts* and *varnish* of Semecarpus Anacardium; called *Marking nut*, from the use made of the juice in marking linen. India. —*Gum* of Spondias Mangifera. India. Resembling *Gum Arabic*. —*Golden Apples*. Spondias dulcis. Tobago.—*Hog-plum bark*. Spondias lutea. B. Guiana. —*Mango; fruits, spirit*, and *wood*. Mangifera Indica. One of the best of tropical fruits. —*Black* or *Munnipoor varnish;* from Melanorrhœa usitatissima. India; and *whips* varnished with the same.—*Sumach*. Dried branches with foliage of Rhus Cotinus, of which 10,837 tons were imported in 1853; used for dyeing black. Sicily. *Wood* of the same.—*N. Am. Sumac; wood, leaves, bark*, and *berries*. Rhus glabra. —*Somacco wood*, of Rhus typhina. Tuscany.—*Japan wax*. Rhus succedanea? —*Cheerongie Oil*. Buchanania latifolia. Madras.

Ord. Amyridaceæ. Amyris Family. (*Case* 22.)—A family especially abounding in balsamic juices, and the leaves are frequently pellucidly dotted. It yields several important products, fragrant gum resins, much used as incenses, and medicinally as stimulants and expectorants. —*Elemi* or *resin*, and *Canarie nuts* and *oil* of Canarium commune. Ceylon and Amboyna. —*Resin* of *Alribe*. Canarium strictum. India. —*Chinese Olives*. Fruit of Canarium Pimela. China. —"*Elemi*," from Mexico. Elaphrium elemiferum.—*Myrrh*. Balsamodendron Myrrha. Arabia. (*Fig*. 9.) —*Balm of Gilead*, or *Balsam of Mecca*. Balsamodendron Gileadense. Syria.—*Guggur*. Balsamodendron Mukul. Scinde. (A similar resin and with the same name is sent from Ceylon, and from Madras).—*Kafal, wood* of Balsamodendron Kafal. Egypt. —*Bdellium* (African). Heudelotia Africana. Senegambia. —*Olibanum*. Boswellia thurifera. India.—*Balsamo*. Bursera gummifera? Venezuela.—*Resin* of Amyris ambrosiaca. Guiana.—*Gum Carana*. Icica altissima. B. Guiana.—*Resin* of Breô branco. Pará. —*Gum resin*, found on the trunk of the large "Gommier tree." Dominica. (Gommier tree is a name given by the French to Bursera gummifera, and to Hedwigia balsamifera, &c.)

Fig. 9.

Balsamodendron Myrrha.

Ord. Leguminosæ. Leguminous Family. (*Cases* 22 to 32 inclusive.)—One of the most extensive of Natural Families, of which the greater proportion, especially those of temperate climates, have papilionaceous flowers, like the pea, the bean, &c.; and all have modifications of that kind of seed-vessel called Legume, even when the flowers are not papilionaceous, as in the Mimosa tribe, the pod in one instance attaining several feet in length! The properties are numerous and various, yielding food, medicine, clothing, dye-stuffs, &c. The Order is conveniently divided into three groups or sections: — 1. *Papilionaceæ*. These have papilionaceous or butterfly flowers, as they are called; and include a large proportion of esculent plants, as the pea and bean. 2. *Cæsalpinieæ*. With conspicuous petals indeed to their flowers, but not papilionaceous; these include many powerful medicinal plants, such as Cassia, &c., and some valuable woods. 3. *Mimoseæ*. With generally small and inconspicuous petals, or none, and with many stamens; the Mimosas and Acacias are examples, and these yield a great deal of mild mucilaginous gum. Their bark is valuable from the quantity of tannin it contains. When the objects are so numerous as in this Family, we can only make a selection for notice in the Guide-Book.

§ I. Papilionaceæ. (*Cases* 22-27.)—*Wild Indigo-root*. Baptisia tinctoria. U. S.—*Sunn*, or *Dhul; fibre* of Crotalaria juncea. India. Extensively substituted for Hemp. Here are *cordage, rope, fishing-lines*, excellent *paper*, &c., made of it. — *Jubblepore-Hemp*. Crotalaria tenuifolia. India. (Other Crotalarias have the same property.)— Espadilla herb. Crotalaria Espadilla. Caraccas. Decoction of Herb refrigerant.—*Lupine Seeds*, Egypt.—*Wood* of *Whin* or *Furze*. Ulex Europæus. A fine specimen, cut in Ireland.—*Baskets*, beautiful in texture and pattern, and *mats*, made of the twigs of Genista virgata. Madeira.—*Broom-tops*. Cytisus scoparius. Much used as an astringent in rustic practice, and a reputed cure for the bite of a mad dog. Eur. — Samples of *seeds* of numerous varieties of *Clover*, of which 214,695 bushels of seed were imported in 1853; *Melilot, Saintfoin, Lucerne*, &c., all from Mr. Lawson's Agricultural Collection.— *Fenugreek seed*. Trigonella Fœnum Græcum. S. Eur.—*Balsam flowers*. Melilotus cærulea. S. Eur. Employed in making the celebrated Chapzigar cheese, to which it gives its flavour and odour.—*Indigo; leaves* and *stems* of Indigofera tinctoria. From this the valuable dye-stuff is obtained. Various samples from E. and W. Indies and Siam. Imports, in 1853, 66,409 cwt. *Spanish* and *Fig* kinds of Indigo.— *Liquorice; root*, and *powder*, and *extract*. Glycyrrhiza glabra. Imports in 1853;—in paste, 209 cwt.; in juice, 9775 cwt. S. Eur. — *Tuberous roots* of Psoralea esculenta, N. Am. — *Wild Coffee Pods*. Sesbania occidentalis. Magdalena, where it is used as Coffee.—*Flaky*, or *Smyrna Tragacanth*. Astragalus verus; and *Vermiform*, or *Moræa Tragacanth*. Astragalus Creticus? Levant.— *Gram*. From seeds of Cicer arietinum. India. Extensively employed as food under that name.—*Chumna ja Bhogra*, or *toasted Gram;* and the same rolled in sugar candy.—*Sayo; cake* made of *Gram* flour, Sesamum oil, and sugar candy. Scinde.—*Peas.* Pisum sativum. In pods (models of 21 vars.) and seed of 55 vars., together with meal of different qualities, and Peas bread. (Lawson's Coll.) Of Peas and Beans, 450,655 quarters were imported to England in 1853. —*Lentils.* Seeds of Ervum Lens (*fig.* 10.) (of which 24,468 bushels were imported in 1853). Much cultivated in N. of Africa and the East; numerous vars. The "*Revalenta*" of the shops. This is nothing but a flour or meal prepared from the seeds of *Lentils,* to which the fabricators give a strangely corrupted name; and, in order, it would appear, to carry the deception further, the advertisements exhibit a tropical scene of lusty negroes cutting down Palm-trees amid Hindu temples, for the preparation of lentil-meal from a humble vetch!—*Tares.* Vicia sativa. Imports of seed in 1853, 17,853 qrs.—*Beans.* Faba vulgaris. Several varieties, and *bean-meal* and *bread.*—*Roots* of Orobus tuberosus. Eur. *Cormeil* of the Scotch Highlanders, who say it allays hunger.—*Pease Earth nuts.* Tubers of Lathyrus tuberosus. Much eaten in Germany, and recommended for general use during the period of the Potato-panic.—*Ground-nuts* and *oil. Pods,* with the *seeds* of Arachis hypogæa; which has the singular property of burying its fruits under ground, where the seeds ripen, and these are used as food, and yield a valuable oil. Africa.—*Shola.* The very soft pithy wood of Æschynomene aspera, and *Sincapore hats* made of it. Used for a variety of purposes, where lightness and softness are required; floats for fishing-nets, &c. Often confounded with the so-called "Rice-paper." India.—*Ordeal Beans* of Old Calabar; a supposed species of Cannavalia.—*Cow itch* (and *pods*). The strong, irritating hairs of the pods of Mucuna pruriens. India.—*Seeds* of Mucuna urens. Jamaica. Drifted (together with the *Cacoon seeds,* Entada gigalobium) by the currents from the West Indies, and washed on shore on the Azores and Loffoden Islands off the coast of Norway and on other northern coasts, directed by the gulf-stream. Of this seed very pretty boxes and ornaments are made.—*Kumur Kus,* or *Butea Kino, Gum Dhao,* from Butea frondosa. The branches of this plant, punctured by a little insect, Coccus Lacca, afford lac of commerce used as dye. India.—*Flowers* of same, which give a beautiful orange dye.—*Tubers* of Apios tuberosa. U. S., where they are eaten; and they also have been tried as a substitute for Potatoes.—*Kidneybeans,* or *Haricots,* and meal. Phaseolus vulgaris. Models of 10 vars. and seeds of 18.— *Bambarra Ground nuts.* Voandezia subterranea. Gambia. Seeds very good; eaten as peas.—*Alcornoque bark.* Bowdichia virgilioides. Venezuela.— *Katchang.*

Fig. 10

ERVUM LENS.

Jepun. Soja hispida, which gives the name to the condiment known as *Soy.*—*Pigeon* or *Cajan Pea.* Cajanus Indicus. E. & W. Ind. Much eaten.—*Crabs' eyes. Seeds* made into necklaces (and *pods*) of Abrus precatorius. India. And the *root* of the same, called *Milhatee,* or *Indian Liquorice* root. Scinde.—*Sissoo wood,* extensively employed in India for the most beautiful furniture, sculpture, &c.—*Red Saunders wood.* Pterocarpus santalinus. Ceylon; a specimen from the Grosvenor ship, wrecked in S. Africa. The tree yields a Gum Dragon. Inner bark of the same, as fibre from Jamaica (cult.).—*Indian Kino* and bark. Pterocarpus erinaceus. Gambia.—*Beejĕsâr wood,* made into a reading-desk of one piece. Pterocarpus Marsupium. Benares.—*Kerim oil.* Pongamia glabra. Chittagong.—*Cabbage-tree bark.* Andira inermis. Jamaica.—*Dogwood bark,* Piscidia Erythrina. Jamaica.—*Rosewood,* for furniture. Triptolemea sp. Imports from Brazil, 5013 tons in 1853, and *Attar* of ditto.—*Tonquin Beans, pod* and *seed,* the fragrant product of Dipteryx odorata. S. Am.—*Eboe nuts* and *oil.* Dipteryx oleifera. Mosquito country.—*Flower buds* of Sophora Japonica; much used as a dye in China and Japan.—*Balsam of Peru,* and basket containing the bottles of the balsam. Myrospermum Pereiræ. Central Am. *Specimens* of the plant, and *seeds* and *wood* of same, and "*spills*" for lighting candles made of the same. Of this Balsam, 34,798 lb. were imported in 1853.—*Balsam of Tolu.* Myrospermum Toluiferum. Columbia.—*Balsamo* or *Estoraque Bark,* and *resin,* and *pastilles* made of same. Myrospermum sp.—*Australian Chestnut.* Castanospermum australe. One of the few esculent seeds of Australia.

§ II. Cæsalpinieæ (*Cases* 27 & 28).—*Logwood.* Hæmatoxylon Campechianum. S. Am., and *extract* of same. Imports of the wood for dye in 1853, 20,752 tons.—*Yellow Nikar nuts.* Seeds of Guilandina Bonducella. S. Am.—*Wood* of *Alleero Morto.* Gymnocladus Canadensis. Tuscany (cult.).—*Sappan Wood.* Cæsalpinia Sappan. India. *Extract* of same, and the same wood from Scinde, as *Bukkum wood.*—*Nicaragua wood.* Cæsalpinia echinata, a valuable dye. Imports, 3318 tons in 1853.—*Brazil wood,* or *Brazilletta.* Cæsalpinia Brasiliensis. S. Am. A dye of much value, probably often confounded with C. crista, also called *Brazil-wood.* Under this name the imports of 1853 were 2658 tons.—*Divi-Divi pods.* Cæsalpinia coriaria. Most powerful astringents, and the pods rich in tannin. 2996 tons were imported in 1853.—*Soap pods* of Chusan. Cæsalpinia.—*Alexandrian Senna.* Cassia acutifolia.—*Tinnevelly Senna.* C. elongata.—*Senna Mekkah* (pods). Cassia obovata. Of Senna, 189,585 lb. were imported in 1853.—*Pods* and *wood* of Cassia Fistula. India.—*Pods* of Cassia Braziliana.—S. Am. Imported as Cassia Fistula.—*Flowers* and *fruit,* and *drawing* of Amherstia nobilis, perhaps the most beautiful tree in nature. Birmah.—*Juruparis* (or devils); musical instruments of the people of the Rio Uaupés (Amazon), made of a tube of Paxiaba Palm stem, enclosed in the bark of Parivoa grandiflora.—*Tamarind pods, preserves,* and *wood;* and *chairs* and *table* made of the very heavy wood (in this room). Tamarindus Indica. Imports in the year 1853, 132,062 bushels, besides 16,661 lb. and at value 382*l.*—*Locust Pods* of *S. Am.* and *Resin,* and *Simiri bark,* of Hymenæa Courbaril. (Here are also several specimens of *gum resins,* "*Animi*" and "*Copals,*" the origin of which is not known.)—*Land-turtles' ladder;* the curiously compressed stems of some Bauhinia.—*Num-num pods,* of Cynometra cauliflora. Malay Islands. —*Balsam of Copaiba.* Copaifera multijuga. S. Am.—*Zapatero* or *Purple-heart wood* of Copaifera pubiflora. Trinidad.—*Mora wood, pods* and *bark* of Mora excelsa. B. Guiana. One of the finest trees and most valuable woods of Guiana and Trinidad.—*Carob pods.* Ceratonia siliqua. S. of Eur. Pulp of the pods sweet and esculent; supposed to be the Locust-tree of Scripture, and often called St. John's Bread. Horses thrive on the pods, and there has been lately a large importation of them into England.—*Cam-wood.* Baphia nitida. Yields a fine red dye for the Bandana handkerchiefs. Sierra Leone. Above 1000 tons are annually imported. *Wallaba bark.* Eperua falcata. B. Guiana. — *Bungo-tree bark* and *resin.* Daniellia thurifera. W. Trop. Afr. — *Wild black Tamarind.* Codarium acutifolium. Sierra Leone.—*Assum Rangie.* Dialium Indum. Java.—*Rosa de montaña,* flowers. Brownea grandiceps. Venezuela. Decoction astringent and used in diarrhœa and dysentery.

§ III. MIMOSEÆ. (*Cases* 28–32).—*Ordeal,* or *Red Water-tree Bark.* Erythrophlœum Guineense. Sierra Leone. The red juice is given in large draughts to the accused; and those who withstand it are declared innocent.—*Gigantic Pods* and *Seeds* of Entada Pursaetha. India, and boxes made of the seeds. Ceylon. Do. of Entada gigalobium. W. Afr. and Jamaica. —*Pods* (curious) of Tetrapleura Thonningii. W. Africa. — *Mexican Gum Arabic,* from Prosopis dulcis.— *Workbag,* made of the seeds of Desmanthus virgatus. Barbadoes.— *Cutch* or *Catechu,* Gum of Acacia Catechu. India. Much used in tanning; and 4389 tons were imported in 1853. — *Mimosa Bark.* Acacia melanoxylon and mollissima, and probably other species of Australia; used in tanning, and imported in extract.—*Myal wood* of Australia, Acacia homalophylla

Fig. 11.

ACACIA ARABICA.

(and probably of other species); remarkable for its delicious and durable odour of violets. There are besides many beautiful samples of *woods* of Australian Acaciæ.—*Gum Arabic:* under this name we have samples of Gums, no doubt from various species of Acacia, but of which the source has probably in few instances, save where there are names attached, been correctly traced as derived from Acacia Arabica (*fig.* 11.); "*Gum Arabic* from the Somali Coast; *Gum Ghaté*, Bombay (the Ghauts)," *E. India Gum*, and *Scinde Gum*, also *wood* of A. Arabica. Scinde. Imports of Gum Arabic in 1853, 61,614 cwt.—*Babool bark*, and *leather* tanned with Babool. — *Scinde Lac.* Acacia Arabica mixed with Zizyphus vulgaris, and *leather* stained with the Scinde Lac.—*Gum Senegal.* Acacia Senegal.—*Gum* also of Acacia Farnesiana (and *flowers* of the same called *Cassia* or *Acacia* flowers, and Attar of do.), and A. rupestris.—*Cape Gum;* from Acacia horrida. —*Gum* of Acacia dealbata. V. D. Land, together with numerous unknown Gums, &c. of this important family.—*Niopo* or *Parica* Pods, and snuff, and apparatus for preparing and taking the snuff. Acacia Niopo. Rio Negro.— *Wood* of *Sabicu.* Lysiloma Sabicu. Benth. A most valuable timber of Cuba.

Ord. Rosaceæ. Rose Family (*Cases* 33–36.)—An extensive Order, of which the Roses are the type, including a large proportion of our esculent fruits; many are astringent, and the stone-fruited kinds abound in hydrocyanic acid. From the kernels of the W. Indian Cherry, Cerasus occidentalis, Noyau is prepared. As in the *Leguminosæ*, so the present Order may be divided into groups or suborders, which by some are raised to the rank of Orders.—§ I. Roseæ. (*Case* 33.)—*Provence Rose petals.* Rosa centifolia. Caucasus; and English *Attar of Roses* from the same, prepared by Messrs. J. Bell and Co. — *Attar of Roses.* From Rosa moschata. Genuine as imported, and adulterated samples. — *Red Rose petals* of Rosa Gallica; and *conserve* of the same. — *Conserve* of *Hips.* Rosa canina. Eur.—*Rose Bedeguars.* Mossy excrescence on rose branches, occasioned by the puncture of an insect, vulgarly "Robin redbreast's pincushions." Eur. —*Purse* made of paste of rose-petals.—*Kousso.* Brayera anthelmintica. Abyssinia. A powerful vermifuge. — *Raspberries preserved.* Rubus Idæus. — *Worsted* dyed with Blackberries. Rubus fruticosus.— *Tormentil, root* of Potentilla Tormentilla. Eur. Very astringent.—*Quillai*, or *Colliguaja bark.* Quillaia saponaria. Chili.—§ II. Drupaceæ, or Stone-fruited Rosaceæ. (*Cases* 33, 34.)—*Bitter Almonds*, Amygdalus communis var. amara; *Sweet Almonds*, do. var. dulcis; *Paper Almonds*, do. var. fragilis. Chiefly from Spain and Portugal. In the second half only of the year 1853, 21,743 cwt. of Sweet Almonds, and in the whole year, 6352 cwt. of Bitter Almonds, were imported.—*Almond oil*, *Essential Oil of Almonds*, *Almond cake* and *Gum* from the wild (Persia and

Barbary) and cultivated Almond; also *Amygdalin.*—The other specimens of *Drupaceæ* consist chiefly of models and preserved specimens of various *Peaches, Camadine,* or preserved *Apricots* in cakes from Damascus, *Nectarines, Plums* (imports of French Plums in 1853, 3986 cwt.; and of Prunes, 36,678 cwt.), *Cherries* (dried, from Portugal); cultivated vars.—*Cherry-gum* and *Cherry-bark.*—*Ornamental* work made of *Peach-stones.*—*Prune-bark* and *Liqueur* of Cerasus occidentalis. Jamaica.—§ III. Pomaceæ. *Apple* and *Pear* group. (*Cases* 35, 36.)—Besides models and specimens of *Apples,* dried American and Normandy (imports of raw in 1853, 355,485 bushels) and *Pears* (imports in 1853 of raw, 48,324 bushels), here will be found the *Quince,* Cydonia vulgaris. Eur. —*Fruit* of *Service-tree.* Pyrus domestica. Eur. — *Medlar.* Mespilus Germanica. Eur.; and the *Loquat,* fruit of Eriobotrya Japonica. China.—§ IV. Sanguisorbaceæ. (*Case* 36.)—A few plants, possessing astringent qualities in the herbage, such as *Common Lady's Mantle,* Alchemilla vulgaris; and *Burnet,* Poterium Sanguisorba. Both from Europe.—§ V. Chrysobalaneæ. (*Case* 36.) — *Coco-plums;* the drupaceous fruit of Chrysobalanus Icaco. Jamaica.—*Grey Plums.* Fruit of Parinarium excelsum. Sierra Leone. *Bitter principle* of the root of Licania amara. Brazil.

Ord. Lythraceæ. Loose-strife Family. (*Case* 36.)— Astringent properties. — *Henna; leaves, powder,* and *fruit,* of Lawsonia inermis. Egypt. Used extensively in the East for dyeing the finger nails. — *Flowers* of Grislea tomentosa. India. A dye. — *Loosestrife.* Lythrum Salicaria. Brit.

Ord. Rhizophoreæ. Mangrove Family. (*Case* 36.) — Mostly shore plants; trees and shrubs of tropical climates, with powerfully astringent bark, yielding a black dye. The branches send down aerial roots, and forests of these are formed at the muddy mouths of rivers, producing malaria. —*Kunro bark* and leather tanned with it. Rhizophora mucronata. Scinde. — *Mangrove bark.* Rhizophora racemosa; *Young Mangrove plant* (and *bark*). The seed germinates while still attached to the parent, and falls down a young plant.

Ord. Combretaceæ. Myrobalan Family. (*Case* 36.)—Tropical trees and shrubs with opposite leaves. Properties astringent, used for tanning and dyeing. — *Myrobalans.* Terminalia Chebula. India. Much used for dyeing. — *Wild-Almond.* Fruit of Terminalia Catappa. India. — *Wood* of Terminalia Belerica. India. Its seed eaten by the natives.—*Gum* of Conocarpus latifolia. India. Its wood is extremely valuable, and there called *button-wood.*— *Sung-sung-hams.* Fruits of Quisqualis indica. China.

First, or Middle Floor (including *Cases* 37 to 69.)

Ord. Myrtaceæ. Myrtle Tribe. (*Cases* 37–40.)—An extensive

family, of warm climates. Trees or shrubs with opposite evergreen leaves, which are pellucidly dotted (indicating the presence of essential oil), and they have an intramarginal vein. Australia is rich in this Family; and among them are some of the loftiest trees in the world, with an enormous thickness of trunk. Several bear edible fruits; some yield spices, whose flavour is due to the aromatic volatile oil.—*Allspice*, or *Pimento*. The fragrant *berries* of Eugenia Pimenta, and *oil* of same; and *basket* made of the fruits. W. Indies. Also called *Jamaica-pepper*.—*Cloves* are, not the flowers nor the fruit, but the *flower-buds* of Caryophyllus aromaticus. Malay Islands; but cultivated now in various tropical climates. Models of a boat, boxes, &c., made of cloves in Amboyna. Imports in 1853, 472,997 lb.—*Myrtle berries*. Myrtus communis. S. of Eur.; and *wood* of the same.—*Cajeput oil*, from Melaleuca minor. Malacca. In great use as antispasmodic, &c. — *Essential oil* of Eucalyptus piperita. W. Australia.—*Wood* of Eucalyptus populifolia. Australia. — *Bark* of a species of Eucalyptus with drawings by the savages of Australia.—*Manna*; from Eucalyptus? V.D.L. — *Australian Kino*. Eucalyptus amygdalina. Australia.—*Jambolins*. Syzigium Jambolanum. Jamaica.—*Pear-shaped Guava*. Psidium pyriferum. Malay Islands.—*Conserve* of *Guava*, and *Jelly*, and *Guava* cheese.—*Apple-shaped Guava*. P. pomiferum.—*Mountain Guava*. P. montanum. Jamaica.—*Fruit* of P. Cattleyanum. — *Rose Apple*. Jambosa vulgaris. — *Malay Apples*. Jambosa Malaccensis. *Rose* var. and *Chili* var. of the same.—*Watery Jambo*. Jambosa aquea. Java. — *Pomegranates*. Punica Granatum. S. of Eur. *Fruit* and *rind*, *flowers*, *root*, *bark*, and *wood*.—*Bogom* or *Bidata*: fruit of Sonneratia acida. Java.

Ord. LECYTHDACEÆ. MONKEY-POT FAMILY. (*Case* 40.) — A small group or Family by many botanists united with the *Myrtles*. The leaves are not dotted; and the fruits are large and in most instances open by a lid transversely; so that the shape is quite that of a pot or caldron with its lid, whence they are called *Monkey pots*. There may be another reason; for the lidless capsule is said to be used for catching monkeys. Sugar is put in, and they are laid in places frequented by monkeys; the opening is small, and it enlarges within, so that when the animal has grasped the sugar with his paw, he is unable to extract it; and the very heavy seed-vessel acts as a clog to him from which he cannot disentangle himself. The genera are all South American. The seeds of some are edible and very good. — *Zabucajo* or *Sapucaia*; fruit and seeds. Lecythis Zabucajo. Brazil. Seeds much eaten. — *Monkey-pots*. The name is given especially to the large heavy kind, Lecythis grandiflora. B. Guiana. —*Tauaré bark*, Lecythis sp. used for caulking ships, and for making the *envelopes* for cigars. Pará. — *Cannon Ball*. Fruit of Couroupita Guianensis. S. Vincent's.—*Castanha*, or *Brazil Nuts*; *Capsule*, and *seeds*, and *bark* of Bertholletia excelsa. Brazil. The seeds most extensively eaten in South America and largely imported into Europe. In the collection are several unknown species of Lecythis.

Ord. BARRINGTONIACEÆ. BARRINGTONIA FAMILY. (*Case* 41.)—*Fruit* of Barringtonia speciosa of the Malay Islands, as well as that of Fœtidia mauritiana, Mauritius, are remarkable in their size or form. *Hindostanee Saleth* or *Slow-match,* of the bark of the *Combu tree.* Careya arborea. Madras.

Ord. HALORAGACEÆ. WATER-CHESTNUT FAMILY. (*Case* 41.)—An aquatic of no importance, except in the following instance.—*Water-Chestnut,* or *Marron d'Eau. Fruit* of Trapa bispinosa, and T. bicornis. S. of Eur. and India. These, which with a little imagination, or a very little assistance from a knife, are very much the shape of a bull's head, are most extensively eaten by some nations, and are very nourishing from their quantity of fecula. Forster says, "They yield a revenue of 12,000*l.* a year to the government of Cashmire; and Moorcroft mentions nearly the same sum, as Runjeet Sing's share, from 96,000 to 128,000 ass-loads of this nut yielded by the Lake of Ooller."

Ord. ONAGRACEÆ. EVENING PRIMROSE FAMILY. (*Case* 41.)—Mucilaginous, some astringent: fruits eaten.—*Scabish Herb.* Œnothera biennis. U. St.—*Bark* of Fuchsia excorticata. N. Zealand.

Ord. CUCURBITACEÆ. GOURD FAMILY. (*Cases* 4–43.)—Herbaceous, succulent, mostly tropical climbing tendrilled plants, of which the Cucumbers and the Gourds are the type. Many bear esculent fruits; and many are acrid and diuretic purgatives. The seeds yield a mild oil.—*Cacoon antidote seeds.* Feuillæa scandens, Jamaica; and *oil* of ditto.—*Bryony root.* Bryonia dioica. Eur.—*Squirting Cucumbers.* Ecbalium agreste. S. of Eur. Remarkable for discharging with great force the seeds and slimy juice through the aperture that is formed by suddenly separating the fruit from the stalk. *Elaterin* derived from do.—*Towel Gourds.* Luffa Ægyptiaca. Jamaica, Africa, &c. The spongy network of the interior may be used as a towel or sponge; and baskets, picture-frames, and other pretty ornamental works are made from it.—*Bottle Gourds.* Lagenaria vulgaris. Tropics. Used for drinking-vessels of various descriptions, owing to the hardness of the shell.—*Snuff-boxes* (Sikkim), and *punch-ladle* (N. Orleans) made of the same.—*Models* and *seeds* of various cultivated Gourds. Cucurbita Pepo.—*Club Gourds.* Probably var. of Cucurbita Pepo; and *Water-vessels* and others, often called Calabashes, from various countries, and some of them beautifully carved.—*Poterion Gourds.* Cucurbita maxima. Models and seeds. Tropics. A *huge vessel* of the same, with a lid, for containing *Poi,* used in travelling by the Sandwich Islanders, is suspended from the ceiling in the principal room of the Old Museum.)—*Orange Gourds.* Cucurbita aurantia. — *Vegetable Marrow.* Cucurbita ovifera. Model and fruits.— *Colocynths, Turkey* and *Mogadore,* peeled and sliced fruit. Cucumis Colocynthis. — *Model* of *fruit* of Cucumis utilissimus. India. Of *Cucumber.* Cucumis sativus. Brazil.—*Melons.* Cucumis Melo. Models. Food and medicine.—*Snake Gourd.* A most remarkable fruit, long and slender, very much resembling a viper or snake, Trichosanthes anguina, in liquid, and models; and *Viper Gourd,* T. colubrina. India.—*Chocho fruits.* Sechium edule. Jamaica. Starch obtained from the same.—*Fruits* of Hodgsonia, &c.

Ord. PAPAYACEÆ. PAPAW FAMILY. (*Case* 43.)—A small Order, of which one plant is worthy of notice, viz. *Papaw fruit.* Carica Papaya. S. Am. Fruit edible. The juice of the entire plant has the property of making old and tough meat tender. Nearly allied to this family is

Ord. PANGIACEÆ. PANGIUM FAMILY. (*Case* 43.)— But here the plants are E. Indian, and poisonous.—*Chaulmoogra seeds* and *fruits.* Gynocardia odorata. Used in cutaneous disorders.—*Fruits* of Hydnocarpus venenatus. E. India. Employed to poison fish in Ceylon; but the fish so captured are unwholesome, and not fit for food.

Fig. 12.

P. CÆRULEA.

Ord. BELVISIEACÆ. NAPOLEONEA FAMILY. (*Case* 43.) — *Flowers* and *fruits,* very curious in structure, of Napoleonea imperialis; named in compliment to the first Emperor Napoleon. W. Africa.

Ord. PASSIFLORACEÆ. PASSION-FLOWER FAMILY. (*Case* 43.)—*Granadilla.* Fruit of Passiflora quadrangularis. This and some other species are favourite tropical fruits.—The curious structure of the flower may be seen in the *model* of the *common blue Passion-flower,* P. cærulea, of Brazil, (*fig.* 12.) where the fancied emblems of our Saviour's Passion (whence the name) are apparent. In the centre are the three nails, — two of the hands, one of the feet (the three stigmas of the pistil); around, the five wounds, indicated by the five anthers of the stamens. The rays of glory (or some say the crown of thorns) are represented by the radiating filaments of the nectary. The floral coverings (ten pieces) are the ten Apostles (two absent—Peter, who denied, and Judas who betrayed, his Master). The wicked hands of his persecutors are seen in the digitated leaves of the plant, and the scourges in the tendrils. Old prints exist with the plant made up of these emblems; and a modern poetess has thus described the discovery and dispersion of this "Flos Passionis:"

"I saw him as he mused one day,
  Beneath a forest bower,
With clasp'd hands stand, and upturn'd eyes,
  Before a *Passion-flower*,
Exclaiming with a fervent joy,
  'I have found the Passion-flower:

"'The Passion of our blessed Lord,
  With all his pangs and pain,
Set forth within a beauteous flower,
  In shape and colours plain.

"'Up, I will forth into the world
  And take this flower with me,
To preach the death of Christ to all,
  As it was preach'd to me.'

"And thus the good old Passion-flower,
  Throughout the world is sent,
To breathe into all Christian hearts
  Its holy sentiment." M. Howitt.

Ord. Portulacaceæ. Purslane Family. (*Case* 43.)—The common *Purslane*. Portulaca oleracea, a potherb, gives the name.—*Melloca* or *Ulluco*. *Tubers* of Ullucus tuberosus. Peru. Esculent and palatable, but no substitute for potato, for which they have been recommended.

Ord. Ficoideæ. Fig-marigold Family. (*Case* 43.)—Of the well-known Cape genus Mesembryanthemum, several are eaten by the natives, and called Hottentot-figs, and others yield soda.—*Capsules* of several, especially M. pomeridianum, &c., are singularly hygrometric; firmly closed in dry weather, spreading like a radiated flower in wet weather, and thus shedding their seeds when the ground is best suited to receive them. *Spætlum*. The thick dried fasciculated roots of a remarkable plant of N.W. America, Lewisia rediviva, collected to a great extent by the Indians, by whom it is extensively used as food on their journeys.

Ord. Cactaceæ. Indian-Fig Family. (*Case* 44.)—The grotesque form of these is familiar to all who visit the Botanic Garden. Some have magnificent flowers; many, esculent fruits. The *Cochineal insect* feeds upon the Opuntia coccinellifer, and may be seen living in the Cactus House of the Garden.—Various *articles* formed of the *fibrous layers* of the Opuntia Tuna.—*Fruits* of some species are preserved in liquor; and such are eaten under the name of *Indian fig* and *Prickly pear*.—*Sections* of the *stems* of some species, exhibiting the oxalate of lime, with which the woody parts abound.—*Preserved* and *eatable pulp* of the fruit of the Cereus giganteus, from New Mexico.—A remarkable specimen of the fruit of Cereus Salmianus, with proliferous fruit.—*Cochineal*, white and black, from Mexico. Of this product of the Opuntia coccinellifer, 10,160 cwt. were imported in 1853.—*Gum*, afforded by an Opuntia in Mexico. *Juice* of the rich colour of cochineal, from the ripe purple fruit of Opuntia horrida. Brit. Honduras. This is suggested as a substitute for the cochineal insect dye.

Ord. GROSSULARIACEÆ. GOOSEBERRY AND CURRANT FAMILY. (*Case* 44.)—*Black* and *red Currants* and *Gooseberries*, preserved. Eur.

Ord. UMBELLIFERÆ. UMBELLIFEROUS FAMILY. (*Case* 45.) — A well-marked and extensive natural Family, including plants that are esculent, poisonous, powerfully medicinal, yielding gum-resins and oily juices. Generally speaking, those that grow in watery and damp places are injurious. — *Balsam-bog.* Bolax glebaria, small specimens from the Falkland Islands, where it forms a striking feature in the scenery. (A gigantic specimen is in the Old Museum.) *Gum-resin* of the same. — *Eryngo root.* Eryngium maritimum. Eur.—*Beela-har Gum-resin* and *seeds,* of Apium maculatum, Loftus'* ms. Persia. — *Umbels,* the rays afford *toothpicks,* of Ammi Visnaga, (Visnaga means a tooth-pick); imported from Africa, and sold at Marseilles.—*Opoponax.* Gum-resin of Opoponax Chironium. —*Ammoniacum Gum-resin.* Dorema ammoniacum. Persia. In lump, and in tears. — *Gum-resin* and *seeds* of Dorema robustum, Loftus' ms. Persia. — *African Ammoniacum.* Ferula Tingitana. Mogadore. —*Stem* of Ferula communis, Candia, the true Ναρθηξ of Dioscorides; *Ferula* of the ancients. The Greeks still call it Nartheca. Tournefort says: "The hollow of the stem is occupied by pith, which, being well dried, takes fire like a match, without injuring the outer portion; and the stem is therefore much used for carrying fire from place to place. Our sailors laid in a store of it. This custom is of the highest antiquity, and may explain a passage in Hesiod, where speaking of the fire that Prometheus stole from heaven, he says that he brought it in a *Ferula,*—the fact being, probably, that Prometheus invented the steel that strikes fire from flint, and used the pith of the Ferula for a match, teaching men how to preserve fire in these stalks." In reality this was a *tinder-box,* now superseded by lucifers. The use of the stems is chiefly for making stools, visiting-seats of the ladies of Amorgos. — *Dill seed.* Anethum graveolens. India. — *Galbanum Gum-resin.* Galbanum officinale. Levant. In lumps and tears. — *Leaves* of Hermas gigantea, used as tinder, &c. S. Afr. — *Hemlock leaves,* and *powder* and *extract.* Conium maculatum. Eur. — *Assafœtida,* gum-resin of Narthex Assafœtida. (*fig.* 13.) In tears. Persia. *Sumbul* or *Musk-root.* Of an unknown plant, sus-

Fig. 13.

NARTHEX ASSAFŒTIDA.

* A. Loftus, Esq. (now carrying out antiquarian researches at Nineveh) presented to this Museum and the British Museum a fine series of Persian gum-resins, accompanied by seeds and specimens, which we trust will throw light upon these valuable yet little known products.

pected to be umbelliferous. Asia.—We must omit, for want of space, and refer to the cases for, the numerous other products of this family named and unnamed. Many are very familiar to our readers, the products as well as the plants; for example, *Parsley*, *Caraways* (imports in 1853, 334,857 lb.), *Anise* and its *oil*, *Fennel*, *Lovage*, *Angelica* (and its candied state), *Parsnep*, *Carrot* (see Drawings on the walls of the top floor), *Chervil*, *Cow Parsnep*, &c. &c.

Ord. Araliaceæ. Ivy Family. (*Case* 45.) — Closely allied to the last order in inflorescence, but generally forming shrubs or small trees. — *American Ginseng* (scarcely different from the famous Chinese Ginseng), Panax quinquefolium. U. S. — *Buah-kadong-dong*. *Fruit* of Panax obtusum. Java. —*American Sarsaparilla*. Root of Aralia nudicaulis. N. Am.—*Rice Paper*. This is the most interesting substance yielded by any plant of this small family; and cases are filled with the preparations, for which, and for a knowledge of the history of the plant, we are chiefly indebted to Sir John Bowring. The *paper*, so called, is the pith of Aralia papyrifera, from Formosa, cut into small sheets; and it is a great article of commerce with the Chinese. Beautiful drawings are made upon it, and the most perfect imitations of flowers made of it. — *Resin* and *fibres* of *Ivy*. Hedera Helix. Brit.

Ord. Cornaceæ. Cornel Tree Family. (*Case* 45.) — *Cornel bark* and *berries* are used medicinally in the U.S. — *Edible fruits* of Benthamia fragifera. Nepal. — Fruits of Cornus mascula, used in Persia for making the celebrated sherbet.

Ord. Loranthaceæ. Misseltoe Family. (*Case* 45.)—The plants of this family are parasitic shrubs, nourished by the trees to which they attach themselves; and the very beautiful *sections* in this case show the union of the parasite with the stock. — Curious *deformed branches* caused by the growth and decay of some large parasitic Loranthi. Guatemala.—Sections of *Misseltoe*, showing its parasitic attachment.

Ord. Caprifoliaceæ. Honey-suckle Family.—(*Case* 45.) Some of the true Honeysuckles entwine a growing stem or sapling so closely as to gird it with a deep spiral line (as in the case of the *whip-handle* made of a sapling Sycamore), giving it a handsome appearance. —To this family also belongs the *Elder*, whose *flowers* and *berries* are used in medicine and household economy.—*Lace* made of *Honey-suckle* fibre. Lonicera Periclymenum. Ireland.

Ord. Rubiaceæ. Madder and Peruvian Bark Family. (*Cases* 24, 25, *and* 26.) — An extensive and important Natural Family, conveniently divided into two natural groups, by some considered Orders: viz. I. *Cinchonaceæ*; shrubs or trees, with opposite leaves and opposite stipules, possessing important medicinal properties; and

II. *Stellatæ*, or *Galiaceæ;* herbaceous plants, with whorled leaves and dyeing properties. We can only mention some of each.—§ I. Cinchonaceæ. *Gum Cumbri*, from Gardenia lucida. Bombay.—*Bark* of Rondeletia versicolor. Veraguas.—*Chayroot.* Hedyotis umbellata. Madras. —*Ipecacuanha root.* Cephaelis Ipecacuanha. (*fig.* 14.) Brazil.—*Striated Ipecacuanha.* Psychotria emetica. S. Am. — *Coffee berries*, in the husk, and freed from the husk. Coffea Arabica. Arabia. Samples from various countries, and some illustrative of Snowdon's

Fig. 14.

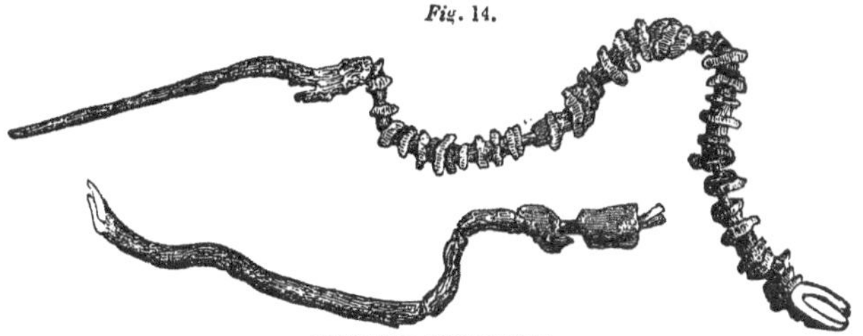

CEPHAELIS IPECACUANHA.

process for cleaning the berry. Of raw coffee, 55,634,733 lb. were imported in 1853. *Coffee-sizer* and *Coffee-peeler* from Ceylon. *Prepared Coffee leaves.* Much used in Sumatra instead of the berry, and containing caffeine; received also from Jamaica.—*Bedula Sutta.* Fibre of Pæderia fœtida. Assam.—*Root, leaves, bark,* and *wood,* of Morinda citrifolia called *Mang : kuda root* in Ceylon. India. Used for dyeing. — *Worsted* dyed with Morinda tinctoria and lac. Sylhet. — For the numerous specimens of *Peruvian Bark,* I must refer to the collections. The best is considered to be the Cinchona Condaminea and C. Calisaya. Of the Peruvian Barks, 13,722 cwt. were imported in 1853; and of Quinine, in half the year, 21,263 oz. —*Sulphate* of *Quinine.*—*Gambier.* Uncaria Gambir (Singapore); and *leaves* and *lozenges.* — *Negro* or *African Peach.* Sarcocephalus esculentus. Sierra Leone.—*Lana, fruit* and *dye.* Genipa Americana. Guiana. —§ II. Stellatæ, or Galiaceæ. *Madder.* Root of Rubia tinctoria. S. of Eur. Also *seed* and *Garancine.* Amount imported in 1853, 111,563 cwt., ground seed; 212,017 cwt. in root, and 22,426 cwt. of Garancine.—*Munjeet.* Root of Rubia cordifolia. India.

Ord. Valerianaceæ. Valerian Family. (*Case* 48.)—Herbs with opposite leaves, strong-scented, anthelmintics and antispasmodics; their effect on cats is very remarkable.— *Valerian root,* and *oil* and *pulvis Valerianæ.* Valeriana officinalis.—*Spikenard.* Supposed to be that of the ancients, and highly prized by them. Root of Nardostachys Jatamansi. India.—*Roots* of Valeriana Celtica.

Ord. Dipsaceæ. Fuller's Teasel Family. (*Case* 48.) —

Fuller's Teasel (and *seed*). The dried flower-heads of Dipsacus fullonum, used for fulling cloth. In 1853 there were imported 14,022,384 Teasel-heads.

Ord. Compositæ. Composite Family. (*Cases* 49–51.) — One of the most extensive and natural Families among plants, very varied in properties; many inert, and many, that have a name as economically useful, are of slight importance. I can only allude to some of the more remarkable, under five heads or groups, into which they are naturally divided: — § I. Vernonieæ (Vernonia the type). *Kenguel. Seed* of Gundelia Tournefortii. Asia Minor. Much used there and in Scinde as Coffee.—*Tinder* from leaves of Andromachia igniaria. Quito. — § II. Eupatoriaceæ. (Type Eupatorium or Hemp Agrimony.) This group has in all countries, especially in the New World, had properties imputed to it of the most extraordinary character, though our best physicians place no confidence in them. Bites of the most venomous snakes, and even cholera, are said to be cured by them.—*Ayapana leaves*. Eupatorium Ayapana. Mauritius. —*Thoroughwort*, or *Boneset*. Eupatorium perfoliatum. U.S.—*Guaco*. Mikania Guaco. S. Am. The celebrated S. American cure for snake-bites.—*Coltsfoot. Flowers* and *leaves*. Tussilago Farfara. Eur.—§ III. Asteroideæ. (Type Aster, or Michaelmas daisy.)—*Herb* of Erigeron Canadensis. N. Am.—*Musk-wood*, very handsome, of Euribia argyrophylla. N.S.W.—*Elecampane. Root* and *powder*. Inula Helenium, Eur.—§ IV. Senecionideæ. (Type Senecio, or Groundsel.) Abounds in plants more or less economic, of which I can only make a small selection; several yield gum-resin.—*Pasto resin*. From an Espeletia. Quito. — *Ramtil seeds*. Guizotia oleifera. Madras. — *Jerusalem Artichokes*. Tubers of Helianthus tuberosus. S. Am. — *Chamomile flowers, extract, oil*, and *green oil*. Anthemis nobilis. Eur. — *Pellitory root*. Anacyclus Pyrethrum. — *Yarrow, herb* and *seed*. Achillea Millefolium. Eur.—*Wormwood, herb* and *extract*. Artemisia Absinthium. Eur. — *Moxa*. Artemisia Chinensis. China. — *Roman Wormwood*. Ambrosia artemisiæfolia. U.S. — § V. Cynareæ. (Type Artichoke and Thistle.) Generally intensely bitter, made bland by cultivation. Roots abounding in gum. Seeds oily and purgative.—*Marygold, flowers* and *seed*. Calendula officinalis. S. of Eur. Used to adulterate Saffron.—*Costus root*. Aucklandia Costus; the Costus of the ancients.—*Blessed Thistle. Herb*. Carduus benedictus. Eur. Used as febrifuge.—*Prepared Thistle down* (Mr. Hyam's).—*Safflower* and *seeds*. Carthamus tinctorius; of which 448,395 cwt. were imported in 1853. India.—*Artichoke, flower*, and *bottoms*, the prepared receptacles of Cynara Scolymus. S. of Eur.—§ VI. Nassauviaceæ (Labiate-flowered).— *Pipitzohoac* acid, together with the plant and root. A most beautiful chemical production, resembling flakes of gold, and said to be acrid, powerfully drastic, with the odour of Valerian. Useful as a dye. It is obtained from the roots of Acourtia rigida. Mexico. (See Hanbury in Pharmaceut. Journ. xv. p. 342.)—§ VII. CICHOREÆ. (Chicory and Lettuce are types.) Properties milky, bitter, astringent and narcotic.— *Chicory. Root, raw* and *prepared, sliced* and *roasted.* Cichorium Intybus. (*fig.* 15.) Eur.—*Scorzonera. Roots* and *seed.* Scorzonera Hispanica. Eur. — *Leaves,* and *extract,* and *Lactucarium* (or *Lettuce Opium*), from the *wild lettuce,* Lactuca virosa; of which, acrid, poisonous, and narcotic as it is, the garden lettuce is a mere cultivated variety. — *Salsify seed.* Tragopogon porrifolius. Eur. — *Dandelion roots,* and *extract,* and *leaves;* and *coffee* prepared from the same. Taraxacum Dens Leonis. Eur. Diuretic and alterative.

Fig. 15.

CICHORIUM INTYBUS.

Ord. CAMPANULACEÆ. HAIR-BELL FAMILY. (*Case* 52.)—Milky and acrid. — *Ramps or Rampions.* Roots of Campanula Rapunculus; cultivated as an article of diet.

Ord. LOBELIACEÆ. LOBELIA FAMILY. (*Case* 52.)—More milky, acrid, and poisonous, than the last family. Sometimes emetic. —*Indian-Tobacco. Herb* of Lobelia inflata. U. S. Much used medicinally in asthmatic complaints.

Ord. GOODENIACEÆ. GOODENIA FAMILY. (*Case* 52.)—*Zaccada pith.* Used as rice-paper pith, and as shola. Scævola Königii. Ceylon.

Ord. ERICACEÆ. HEATH FAMILY. (*Case* 52.)—Small trees or shrubs, remarkable for the beauty of their flowers, rather than for any very useful properties. They are generally astringent; the common Ling is used for dyeing. Some have poisonous, narcotic qualities, as Rhododendron and Azalea. Some fruits are eatable.—*Uva Ursi. Leaves* and *berries* of Arctostaphylos Uva Ursi. Eur. and N. Am. Used as an astringent medicine.—*Shallon. Berries,* and *cake* or *bread* of Gaultheria Shallon. Much eaten in N. W. Am.—*Poisonous honey* of Azalea Pontica from Erzeroum.—*Wood,* and *Yak saddle* made of Rhododendron arboreum. Tibet. *Jelly,* for the table, made of the petals of the same. Himala.—*Leaves* of Rhododendron chrysanthum. Siberia. Very poisonous.—*Dried tuft* of Rhododendron nivale; from Kinchinjunga, E. Himala. elev. 17,500 feet above the sea. Figures of the *Sikkim rhododendrons* are suspended on walls of this floor in Room No. 6. Many Rhododendrons contain gum-resin.—*N. American Winter-green; berries* and *oil* of same. Gaultheria procumbens.—*Yellow worsted* dyed with common *Ling,* Calluna vulgaris. Scotland. —Nearly allied to Ericaceæ, are *Vacciniaceæ* (*Case* 53.), of which we have *Cranberries*, (Vaccinium Oxycoccos), and *greater Cranberries*, V. macrocarpum (other Vaccinia have also esculent fruits); and *Pyrolaceæ*, to which belongs *Pipsissewa*, Chimaphila umbellata. Tonic and narcotic. U. S.

Ord. Styraceæ. Storax Family. (*Case* 53.)—Properties stimulant, aromatic, and fragrant; yielding balsamic gums.—*Benzoin, lump*, and in *tears*. Styrax Benzoin. Sumatra, Siam.—*Common Storax* (Styrax Calamita,) and *wood* and *bark*, and *liquid Storax*. Styrax officinale. Syria and Arabia.—*Bootan cloth*, dyed with a Symplocos. Sikkim.

Ord. Ebenaceæ. Ebony Family. (*Cases* 53 *and* 54.)—Remarkable for the hardness of the wood. Some bear esculent fruit.—*Ceylon Ebony*. The very hard black wood of Diospyros melanoxylon, various manufactured articles. — *Calamanda wood*, highly valued, and boxes made from this beautiful material by the Cingalese. Diospyros hirsuta. Ceylon.—*Edible fruits* of several Indian species of Diospyros.

Ord. Aquifoliaceæ. Holly Family. (*Case* 54.)—Evergreen trees, generally astringent and tonic. — *Common Holly, wood*, and *Bird-lime* made of the bark. Ilex Aquifolium. Eur. — *Maté* or *Paraguay Tea*. *Leaves* of Ilex Paraguensis: used as tea over a great portion of S. America, where the consumption is enormous. Large bag of Ox-hide; and an ant-eater's skin stuffed, with the maté, packed for export, presented by Capt. G. F. Morice, R.N. (these, for want of space here, are to be seen in the Old Museum, Room 8, *Case* n. 82). Brazil. *Maté cups* and *tubes* used in drinking the infusion.—*Leaves* of Ilex vomitoria. S. U. States. A strong decoction of them is always drunk by the Creek Indians at the opening of their councils; and they send annually to the seacoast for a supply of the leaves.

Ord. Sapotaceæ. Sapodilla Family. (*Cases* 55–57.) — Astringent, tonic, and febrifuges. Many contain viscid milky juices, somewhat analogous to Caoutchouc; others, oil and fatty matter. — *Butter* of *Madhuca-tree*, together with *seeds* from the same. Bassia butyracea. India. Used as butter. — *Shea butter* and leaves and fruit. Bassia Parkii. Sierra Leone. — *Nancel oil*. Bassia latifolia. Canara. *Mahwa flowers* of the same. — *Gutta Percha*. The inspissated juice of Isonandra Gutta. Singapore and Malay Islands. An entire case exhibits, besides a dried specimen of the Plant itself, in a frame made of its juice, the raw material in lumps (of which 21,792 cwt. were imported in 1853), and some of the numerous articles that are manufactured of this wonderful substance, unknown in European commerce till 1845. These lumps or cakes are a foot or more in diameter each way. At the manufactory they are cut into slices, cleaned from stones and other impurities, and boiled in waste steam from the engine. Reduced to a somewhat uniform consistency, the mass is put into an Iron Box or "Teaser," in which an iron cylinder, with teeth, rapidly revolves and tears it into shreds, throwing it into vats of cold water. There the Gutta percha floats at the top, and the impurities sink to the bottom. It is then transferred to tanks of boiling water, removed into boxes and "kneaded," and is then in a state to incorporate with other substances, such as India rubber, which gives it a greater degree of elasticity. It is now rolled out into sheets or tubes, or moulded into various forms and for various purposes. Some of its applications are ingeniously alluded to in the following lines, written by an anonymous visitor to the great Gutta Percha establishment in the Wharf-road, City-road:—

1. "My Parent died when I leap'd from her side To fill mankind with wonder,	1. The gutta percha trees are tapped and they then die.
2. And now I abound, in the wide world around, The green sward above and under.	2. Used above and below ground.
3. I hold the flower in the sunny bower;	3. Flower-pots.
4. I shelter the dead in their graves.	4. Lining for coffins.
5. I circle the hair of the maiden fair;	5. Bonnet-caps.
6. And bid defiance to knaves.	6. Policemen's ditto.
7. The miser his gold often gives me to hold;	7. Money bowls.
8. I aid to extinguish the fire.	8. Water buckets and engine-pipes.
9. I am chased o'er the green, when the schoolboy is seen;	9. Cricket balls.
10. I wait at the toper's desire;	10. Mugs.
11. I ride on the wave, the sailor to save, When he shrieketh aloud in despair;	11. Life-buoys and boats.
12. I whirl the machine, whose arms dimly seen, Hiss as they fly through the air.	12. Machine-driving belts.
13. I have been tried and am cast, with felons at last;	13. Indestructible vessels for the use of prisoners.
14. I am balm to the wounded and torn.	14. Balsam for slight wounds, instead of sticking plaster.
15. I rival the oak, (16.) the tell-tale I cloak;	15. Ornamental mouldings. (16.) Coating of the Telegraph wires.
17. I am fashion'd as high and low born.	17. Medallions and casts of celebrated and notorious persons.
18. I constantly mind the sightless blind.	18. Cord for window blinds.
19. Many garments my long arms bear;	19. Clothes-lines.
20. By the sick man's bed, (21.) by the ship's mast-head— In various forms I am there.	20. Utensils for sleeping apartments. (21.) Cordage and speaking-tubes.
22. Deep in the earth, though unseen, is my worth; I faithfully serve mankind.	22. Pipes for draining, &c.
23. I hear the whisper of the softest lisper;	23. Acoustic tubes.
24. And hold that which traceth the mind!	24. Inkstands.
25. When the emigrant lands on far-off strands, Perchance he treadeth on me.	25. Soles of shoes.
26. On the rich man's table, (27.) in the horse's stable, My forms you may frequently see! Now I challenge your mind my secret to find,	26. Ornamental dishes. (27.) Buckets and harness.
28. Though I travel along by your bed.	28. Noiseless curtain rings.
29. I come from the south; (30.) I may dwell in your mouth;	29. From Singapore, &c. (30.) For filling decayed teeth.
31. Or may rest on the top of your head."	31. Sou'-wester hats.

—*Mamay de Cartagena.* Achras mammosa. N. Grenada.—*Naseberry* or *Sappodilla;* the fruit of *Achras Sapota,* and a *spirit* from same. W. Ind. Fruit very agreeable after being kept, like the medlar.—*Star Apples.* Chrysophyllum Cainito. Jamaica. — *Argan. Branches* in *flower* and *fruit,* and *oil* of Argania Sideroxylon. Morocco. A Tree of great importance to the Moors. The pulpy fruits are eaten by all ruminating animals; and on chewing the cud, the hard seeds are ejected, from which a valuable oil is extracted, almost equal to olive oil. Wood hard and useful. — *Oil* from flowers of Mimusops Elengi. Amboyna. — *Buah Sow. Fruits* of Mimusops Kauki. Java. *Monesia* bark. Chrysophyllum.

Ord. Jasminaceæ. Jasmine Family. (*Case* 57.) — Roots bitter. Flowers highly fragrant, and yield aromatic oils.— *Flowers* of Nyctanthes Arbor-tristis. India.— *Oil* from Flowers of Jasminum Sambac.

Ord. Oleaceæ. Olive Family. (*Cases* 58 *and* 59.)—Bitter, tonic, and astringent, and yield fixed oil.—*Olives, Wood, Oil,* and *Spoon* made of the wood, *rosary* made of the wood, and also of the seeds from Mt. Tabor and from Jerusalem (Mount of Olives). Olea Europæa. S. of Eur. and Palestine. Oil imported in 1853, 10,102 tons. — *Manna, flake,* and *Mannite,* from Fraxinus Ornus. S. of Europe. — *Chinese Insect Wax,* "*Pela,*" (together with *the insect,* Flata limbata,) considered to be due to the insect feeding on a plant of this family, Ligustrum lucidum? The wax is imported from China, and esteemed valuable. *Candles made* of *tallow,* dipped in *Pela.*—*Iron-wood* of Norfolk Island. Olea apetala.

Ord. Salvadoraceæ. Mustard-tree Family. (*Case* 57.) — Blocks of the *wood* of the *Mustard-tree* of Scripture, according to Dr. Royle, Salvadora Persica. The East. — *Wood* of another species of the genus, S. oleoides. Scinde.

Ord. Asclepiadaceæ. Asclepias Family. (*Case* 59.) —Abounds in milky, generally acrid, and poisonous juices, often also in beautiful fibre.—*Indian-Sarsaparilla.* Hemidesmus Indicus. India. Sold in London as "Rad. smilax aspera."—*Indian Hemp-root.* Asclepias incarnata. U.S.—*Pleurisy root.* Asclepias tuberosa. U.S. — *Paper* made of an Asclepias. India.—*Muslin* made of an Asclepias. Syria. — *Band* made of the fibre of Leptadenia Jacquemontii. Scinde. — *Fibre* of Marsdenia tenacissima, of Calotropis procera, C. gigantea, and of other Asclepiadaceous plants.—*Silky substance* of the seeds of Calotropis gigantea. India. — *Ascidia,* or *pitchers* of Dischidia Rafflesiana. Java.

Ord. Apocynaceæ. Dogbane Family. (*Case* 60.) — Some are milky, many poisonous; others yield edible fruits.—*Tanghin fruit.* Tanghinia veneniflua. Madagascar. The poison ordeal is employed with persons suspected of crime; but so prepared by the priests, that it becomes injurious or otherwise, at the pleasure of the administrator.—*Paddle-wood* of Aspidosperma excelsum. Used for paddles, remarkable for the sinuosities of the wood (large block on the ground in the Room, under a window).—*Gutta Jintiwan.* Urceola elastica. Malay Islands. One of the Caoutchoucs.—*Seeds* of Wrightia antidysenterica. India.—*Oil* from roots of Alyxia stellata. Java.

Ord. LOGANIACEÆ. STRYCHNOS FAMILY. (*Case* 60.)—Acrid, astringent, very poisonous.—*Indian-Pink.* Spigelia Marilandica. U. S.—*St. Ignatius' Bean.* Ignatia amara. Said to be a cure for cholera in India, but a dangerous remedy.—*Nux vomica, Capsules* and *seeds, bark* (called *false Angostura*), *wood,* and pure *Strychnine.* Strychnos Nux vomica. E. Ind.—*Ourari* or *Wourali poison* (paste), and *wood* and *fruit.* Strychnos toxifera. Guiana.—*Clearing nut.* Strychnos potatorum. India. Used for purifying water.—*Snake-wood.* Strychnos colubrina.

Ord. GENTIANEÆ. GENTIAN FAMILY. (*Case* 61.)—Bitter in an eminent degree, and tonics.—*Gentian-root.* Gentiana lutea. S. of Eur.—*British Gentian-root* and *powder.* Gentiana campestris. Eur.—*Buck-bean, leaves* and *fruit.* Menyanthes trifoliata. Eur. and N. Am.—*Common Centaury.* Erythræa Centaurium.—*Chirita* or *Chirayetta.* Ophelia Chirita. India.

Ord. BIGNONIACEÆ. TRUMPET-FLOWER FAMILY. (*Case* 61.)—Handsome climbing plants, with curious, often large, pod-like seed-vessels, and beautiful winged seeds, but of little importance economically.—*Chica, Leaves* and *pigment.* Bignonia Chica. S. Am.

Ord. PEDALIACEÆ. SESAMUM FAMILY. (*Case* 61.).—*Teel, Gingelly* or *Benne seeds.* Yield a copious, pure oil. Sesamum orientale. India.—*Fruits* of Martynia proboscidea. India; and of the *Grapple plant.* Harpagophytum procumbens. Cape.

Ord. CRESCENTIACEÆ. CALABASH FAMILY. (*Case* 61.).—A small Order, remarkable for the size and hard shells of the fruits, rendering them available for drinking- and other -vessels, bowls, or calabashes; often beautifully carved by the natives.—*Various fruits* and *plain* and *ornamented calabashes*; chiefly of Crescentia Cujute. W. Ind.—*Fruits* of Kigelia pinnata (Mauritius), and of K. Africana. Trop. Afr.

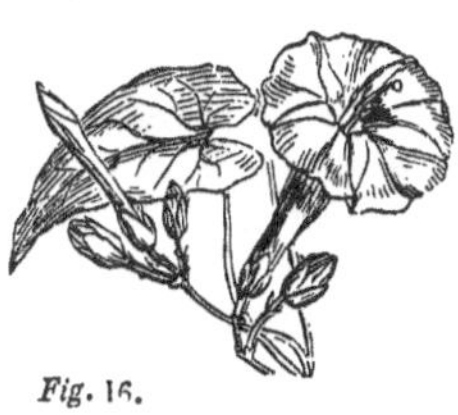

Fig. 16.
EXOGONIUM PURGA.

Ord. CONVOLVULACEÆ. BINDWEED FAMILY. (*Cases* 61, 62.)—Mostly climbing plants; the roots purgative.—*Jalap,* root and powder. Exogonium purga. (*Fig.* 16.) Mexico; of which 82,345 lb. were imported in 1853.—*Mechoacan root.* Batatas Jalapa. Mexico. By some considered the true Jalap.—*Scammony*. Convolvulus Scammonia. Levant. Imports in 1853, 9704 lb.—*Sweet Potatoes*, or *Batatas; roots* and *models* of same. Convolvulus Batatas.—*Lace* made of the *fibre* of Convolvulus arvensis. Ireland.—*Rhodium Oil*, from Convolvulus scoparius.—*Turbith root*. Ipomœa Turpethum. India.

Ord. Cuscutaceæ. Dodder Family. (*Case* 62.)—*Dodder* (injurious to field crops), *seed* and *oil*. Cuscuta Europæa, &c. Eur.

Ord. Boraginaceæ. Borage Family. (*Case* 62.)—Mucilaginous and emollient, often yielding dyes in their root, as *Alkanet root*. Anchusa tinctoria. S. Eur.—*Roots* of Cynoglossum officinale, or *Hound's tongue*. France.—*Comfrey root*, Symphytum officinale; and *Gromwell seeds*, Lithospermum officinale. France.

Ord. Cordiaceæ. Cordia Family. (*Case* 62.)—*Fibre* of Cordia Sebestena, C. macrophylla, and C. Gerascanthus. Jamaica.—Rope made of Cordia myxa. Ceylon.—*Seeds* of Cordia grandiflora. India.

Ord. Solanaceæ. Nightshade Family (*Cases* 62 *and* 63.).—Properties, in some narcotic and poisonous, in others acrid and irritant, owing to an oleaginous substance, called Capsicine.—*Dulcamara*, or *Bitter-sweet*. Solanum Dulcamara. Berries poisonous.—*Potatoes*. *Tubers* of Solanum tuberosum. Peru and Chili. *Native tubers* (small) and various-named cultivated ones. *Dried sliced tubers*, *starch* from sound and diseased tubers. *Floating* residue of the potato, after the starch is deposited. *Gum substitute* (5 sorts). Among the Potatoes will be seen a small stand, on which Professor Henslow has placed models of three potatoes, which together weighed exactly *one pound**; and, in front of these, he has given the respective amounts of the four principal ingredients of which this pound is composed:—

Water - -	74 per cent, or	11 oz.	16 dwt.	19·2 gr.
Vegetable fibre -	16	,,	1	5	14·4
Starch - -	8	,,	2	11	4·8
Casein - -	2	,,	0	6	9·6

The amount of carbon in the *three solid* ingredients is also given apart, being 1 oz. 17 dwt. 3 gr. In a small glass tube is likewise shown the small quantity of *inorganic* matter contained in a pound of potatoes, but not regarded in the above approximate analysis. This amounts to 2 dwt. 54·7 gr., and constitutes the *ashes*, when the vegetable matter is consumed by heat. It supplies materials for bone. Large as is the proportion of water, it is not more than exists in a pound of flesh. The fibre is probably indigestible. The starch affords an abundant supply of carbon, essential to the process of respiration, without drawing upon the casein, which, being of the same

* The average consumption of this food in Ireland is stated to be at the rate of about 11 lb. a day for each individual.

composition as flesh, is the only ingredient in the potato employed in restoring this substance; in other words, of strictly *nourishing* the body. The pulpy mass of vegetable tissue obtained by crushing potatoes for the extraction of starch becomes hard when dried, and then forms an excellent material for the manufacture of snuff-, and other horn like, -boxes.—*Tomatos.* Lycopersicum esculentum. *Capsicums* and *Chillies.* These are varieties or species of Lycopersicum, cultivated in various countries.—*Winter Cherries* or *Cape-gooseberries.* Physalis Peruviana. Excellent fruit for the dessert. S. Am., &c.—*Puneer berries.* Puneeria coagulans. Used for coagulating milk in Scinde.

Ord. ATROPACEÆ. BELLADONNA FAMILY. (*Cases* 64. *and* 65.)—Properties similar to the last Order, but the narcotic and poisonous principles are more highly developed.—*Stramonium; leaves, powder, capsules, seeds,* and *Extract* of Datura Stramonium. Eur.—*Henbane; seed, leaves, Cigars, powder,* and *Extract.* Hyoscyamus niger. Eur. — *Belladonna; leaves, Cigars, berries, Extract,* and *root,* and *Atropine.* Atropa Belladonna. Eur. — *Mandrake root* and *fruit.* Mandragora officinarum. S. Eur. — *Tobacco.* Nicotiana Tabacum, &c. Persia, &c. Thanks to our mercantile friends, this collection is so rich in kinds, species and varieties, *Cigars* * and *Snuff,* that we can only refer to the named samples in *Case* No. 65. of this Room, and upon the adjacent walls outside the case. The imports of Tobacco in 1853 are given at 40,670,032 lb.; of Cigars at 4,308,164 lb.; and of Snuff at 3923 lb. *Tobacco seed Oil.* Tanjore.

Ord. SCROPHULARIACEÆ. FIGWORT FAMILY. (*Case* 66.)—Acrid, and slightly bitter, without aroma. Some acrid, and some sedative, others poisonous. Several are parasitic on the roots of other plants. A few only are employed in the Materia Medica. Principally — *Digitalis* or *Foxglove.* Digitalis purpurea. Leaves and powder.—*Eyebright,* formerly in much use as *Euphrasy.* Euphrasia officinalis.—*Gratiola, herb.* Gratiola officinalis.—*Figwort.* Scrophularia nodosa. All Eur. — *Roots* of Picrorrhiza Kurroa. Kamaon. — *Balmony herb.* Chelone glabra. U. States.—*Mullein.* Verbascum Thapsus. Brit. — *Toad-flax.* Linaria vulgaris. Brit.

Ord. OROBANCHACEÆ. BROOM-RAPE FAMILY. (*Case* 66.)—*Specimen* of Orobanche, showing its *parasitism* on clover.

* The most extensive manufactory of cigars in the world, perhaps, is at Manilla, where Captain Wilkes, U. S. Navy, inspected it; and we must make room in a note, for an extract relating his visit:—There are two establishments. The one we inspected had sentinels at all the gates, and 8000 females were employed! Throughout the whole of the vast apartments (all kept in the neatest possible order) tables are arranged, ten feet long and three wide, at each of which fifteen women are seated, with small piles of tobacco before them. Each female makes 200 cigars a day. "It is a strange sight," he says, "to witness the ingress and egress of these women; and probably the world cannot elsewhere exhibit so large a number of ugly ones. In age they vary from 15 to 45. The whole number of persons employed in the two manufactories is about 15,000, and all are searched twice a day in case they should be guilty of pilfering. The revenue to the country derived therefrom is said to be 500,000 dollars."

Ord. LABIATÆ. LABIATE FAMILY. (*Cases* 66, 67.)—An extensive Order, readily distinguished by the 4-angled stems, opposite leaves, and labiate flowers generally arranged in whorls, and very deeply 4-lobed ovary. It abounds in reservoirs of volatile oils, and is fragrant and aromatic, possessing no dangerous qualities. Many yield stearoptin, resembling camphor.—Among them we here find *Basil-seed.* Ocymum Basilicum. S. Eur.—*Lavander ; flowers,* and *seed,* and *oil.* Lavandula vera.—*Patchouli ; herb,* and *essence,* and *oil,* long held in great reputation as a perfume; it is this which gives the peculiar odour to Indian or Chinese Inks. Pogostemon Patchouli. India.—*Spearmint* and *oil* of Mentha viridis.—*Peppermint* and *oil.* Mentha piperita. — *Pennyroyal* and *oil.* Mentha Pulegium.—*Sage.* Salvia officinalis.—*Clary.* Salvia Sclarea. Seed and flowers.—*Rosemary* and *oil.* Rosmarinus officinalis.—*Sweet Marjoram* and *oil.* Origanum Marjorana. — *Thyme* and *oil.* Thymus Serpyllum.—*Hyssop.* Hyssopus officinalis.—*Balm.* Melissa officinalis.—*Horehound.* Marrubium vulgare. All European.— *Skull-cap.* Scutellaria lateriflora. U. States.

Ord. VERBENACEÆ. VERVAIN FAMILY. (*Cases* 67–69.)—No very marked properties. Some are fragrant and aromatic. Teak is one of our most valuable ship-timbers.—*Verbena, herb.* Aloysia citriodora; abounds in fragrant essential oil, as do several others of the same genus from S. America. —*Vervainia.* Stachytarpheta jamaicensis. Jamaica. — *Teak* of India. *Wood* of Tectona grandis. (Much used in our navy.) *Idols,* &c., carved from it. The quantity of teak imported in 1853 was 14,931 loads; but this probably includes African Teak (Oldfieldia Africana). — *Courida Bark.* Avicennia nitida. S. Am. Used in tanning.—*Agno Casto, wood.* Vitex Agnus castus. Tuscany. Fruits acrid; but used as a remedy for colic at Smyrna. — *Fiddle-wood,* Citharexylon melanocardium. Jamaica.

Ord. ACANTHACEÆ. ACANTHUS FAMILY. (*Case* 69.)— Mucilaginous and bitter. — *Kreote.* Justicia paniculata. Madras.— *Herb* of Rhytiglossa pectoralis. Guiana.

Ord. PLUMBAGINEÆ. LEAD-WORT FAMILY. (*Case* 69.) Slightly acrid and tonic. — *Lead-wort root.* Plumbago Europæa.— *Stalks* of Plumbago Zeylanica. India.

Ord. PRIMULACEÆ. PRIMROSE FAMILY. (*Case* 69.).—*Cowslip flowers.* Sedative. Primula veris. Eur. A pleasant wine is often made of them. — *Arthranite root.* Cyclamen Europæum. — *Herb Pimpernel.* Anagallis arvensis. Brit.

GROUND FLOOR (including *Cases* 70 to 100).

Ord. NYCTAGINEÆ. MARVEL OF PERU FAMILY. (*Case* 70.) —Purgative; as *Root* of Mirabilis Jalapa. Mexico. Long supposed to be the true Jalap; so named as coming from near Xalapa.

Ord. CHENOPODIACEÆ. GOOSE-FOOT FAMILY. (*Case* 70.)—Chiefly potherbs.—*Stinking Goosefoot.* Chenopodium olidum. Eur.—*Quinoa seed, black* and *white.* Chenopodium Quinoa. Much used as food in Chili and Peru, by all grades of people.—*Wormseed.* Chenopodium anthelminticum. N. Am.—*Beet-root*, or *Mangold Wurtzel.* Beta vulgaris. *Models, Roots, dried, roasted,* and *ground* (for coffee), and *sugar.* Much sugar is made from Beet, especially in France and lately in Ireland; it is finer, but less sweet than that from cane.—*Sujjee-Khar*, or *Koree Khat.* Impure Carbonate of Soda from *Salt-worts,* species of Salsola. Scinde. (Large quantities are made in Spain and Portugal from the plants of this family, that grow in saline situations.)—*False Samphire.* Salicornia herbacea. Eur. Often pickled and sold and eaten as true samphire.

Ord. PHYTOLACCACEÆ. POKE-WEED FAMILY. (*Case* 70.)—*Garget,* or *Poke-weed; Root* and *Berries.* Phytolacca decandra. N. Am. The former used in medicine; the latter for staining wine; and hence the plant has been imported into and grown in Europe.—The peculiarly soft and rapidly growing wood of Picrunia dioica, of S. Brazil, called *Bellasombra,* much planted to afford shade in walks.

Ord. POLYGONACEÆ. BUCKWHEAT FAMILY. (*Cases* 70 & 71.)—Astringent, and often very acid. The seed farinaceous, several kinds used medicinally, and as food.—*Water Dockroots,* Rumex Hydrolapathum; and *Yellow Dockroots,* Rumex Patientia. Eur.—*Bistort; root* and *seed,* Polygonum Bistorta. Eur.—*Buckwheat; seeds, flour,* and *bread.* Polygonum Fagopyrum. Much used as bread in Germany, and for feeding fowls and game in England.—*Rhubarb; root, seed,* and *powder;* various sorts and qualities; but the exact species producing the best medicinal roots is, in very few cases, accurately known. *Leaf-stalks* of same, for making tarts and wine; *models.*—*Black Worsted,* dyed with *Sorrel* roots, Rumex acetosa. Eur.

Ord. LAURACEÆ. LAUREL FAMILY. (*Cases* 71–73.)—Mostly aromatic and fragrant, yielding fixed volatile oils, camphor, and tonic bark.—*Cinnamon; bark, oil* of *bark,* and of *leaf.* Cinnamomum Zeylanicum. Ceylon. Very fine specimens of *bark, oils,* and *implements* for peeling. Imports in 1853, 220,586 lb.—*Cassia,* or *Cassia lignea; bark, oil, wild* or *Java Cassia, Oan-Qui-Pe* or *thick Cassia* (from China), and *Cassia-buds.* Cinnamomum Cassia. India.—*Camphor; sublimed,* and *wood,* with *Camphor sublimed in the bottle.* Camphora officinarum. China.—*Beeberu bark* and *seeds.* Nectandra Rodiæi. Powerful febrifuge. B. Guiana; and *wood* of the same called *Greenheart.*—*Avocado Pear.* Persea gratissima. W. Indies.—*Laurel oil.* Oreodaphne oleifera. S.Am. B. Guiana.—*Sassafras; wood* and *oil, bark* and *chips.* Sassafras officinale. N. Am.—*Spice-* or *fever-bark leaves.* Benzoin odoriferum. U. S.—*Sweet Bay* or *Victor's Laurel; berries, oil,* and *wood.* Laurus nobilis. S. Eur.—*Ackawai Nutmegs.* Acrodiclidium Camara. Guiana. — *Tawa seeds.* Laurus Tawa. N. Zealand. — *Oil* of Laurus Canariensis. Madeira. — *Ravansara nuts.* Agathophyllum aromaticum. Madag. — *Pichurim* beans. Nectandra Pichury major. Brazil.— *Clove-bark.* Dicypellium caryophyllatum. Pará.

Ord. Atherospermaceæ. Plume-Nutmeg Family. (*Case* 74.) *Sassafras bark* of Tasmania. Atherosperma moschatum.

Ord. Myristicaceæ. Nutmeg Family. (*Case* 74.) — A small Order, of which the Nutmeg is the individual most worthy of attention. The fruit is fleshy; within the fleshy substance is the Mace, within that again the Nutmeg.—*Nutmegs, wild* and *cultivated; entire fruits, mace* and *nut; attar, oil, fluid* and *concrete,* and *jelly.* Myristica moschata, (or M. officinalis). Malay Islands. Imports of Nutmegs in 1853, 300,563 lb., and of Mace, 83,022 lb. — *Fruit; nut,* and *fatty matter,* obtained from Myristica sebifera. S. Am. and *fat* from M. Otoba. Antioquia.

Ord. Proteaceæ. Protea Family. (*Cases* 74–76.)—No peculiar properties. The large *cone-like fruits* of Banksia and Dryandra afford food to parrots and other hard-billed birds in Australia. The beautiful Leucadendron argenteum is the common fire-wood of the Cape; and Guevina Avellana yields the edible nuts of Chili, called *Avellana.* Many of the *woods* exhibited of this family are very beautiful.

Ord. Eleagneæ. Oleaster Family. (*Case* 76.) — *Fruit* of *Sea-Buckthorn.* Hippophae rhamnoides; it makes a good preserve. Eur. — *Gum* of Elæagnus angustifolus. Scinde.— *Trebizond-dates.* Elæagnus orientalis.

Fig. 17.

Lagetta lintearia.

Ord. Penæaceæ. Sarcocol Family. (*Case* 76.)—*Sarcocol, Gojan, Anziroot; Gum-resin* so called from Penæa Sarcocolla. Cape.

Ord. Thymelaceæ. Spurge-Laurel Family. (*Case* 77.) — Acrid and irritant, sometimes narcotic, sometimes acting as poison; abounding in fibre in the inner bark.—*Mezereon root, stems* and *bark.* Daphne Mezereum. Eur.—*Spurge-laurel, leaves* and *root.* Daphne Laureola. Eur. —*Inner bark,* and *dried pulp,* and *paper* made of Daphne cannabina, and D. Gardneri. E. Indies.—"*Burn-nose,*" *bark* and *ropes.* Daphne tinifolia. Jamaica.—Garon bark. Daphne Gnidium. Eur. — *Jamaica Lace-bark;* the lace-like layers of the inner bark of Lagetta lintearia (*Fig.* 17) of Jamaica: a great variety of useful as well as ornamental articles made therefrom, as a substitute for lace; also *whips, thongs, and handles*, &c. &c.—*Leather-wood, Bark.* Dirca palustris. U. S.—*Poovana oil* from Sarcostigma Kleinii. Tinevelly.—*Leather-wood tree* bark. Dirca palustris. U. S.—*Fruits* of *Hernandia ovigera.* W. Ind.

Ord. Aquilariaceæ. Eagle-wood Family. (*Case* 78.)—*Lignum Aloes, Eagle-wood; wood, Rosaries* made therefrom, and *oil* (called *Uggur*). Aquilaria Agallocha. Asia. Exceedingly fragrant, and supposed to be "Aloes" and "Lign-aloes" of Scripture.

Ord. Santalaceæ. Sandal-wood Family. (*Case* 78.)—Well known from the fragrance of the chief species.—*Sandal-wood.* Santalum album. India and Pacific Islands (probably different species). *Wood, chips, oil, sticks*, and *carved box.*—*Oil* of *Sandal-wood*, from Santalum myrtifolium. India.—*Ro-cha.* Spurious Tea of Kunawar. an *Osyris.*

Ord. Aristolochiaceæ. Birth-wort Family. (*Case* 78.)—Bitter, tonic, and stimulant.—*Asarabacca.* Asarum Europæum; "*British herb-snuff.*" *Leaves* and *powder.* Eur.—*Snake-root.* Asarum Canadense. U. S.—*Birth-wort Root.* Aristolochia Clematitis. Eur.—*Serpent-root.* Aristolochia serpentaria. U. S.—*Round Aristolochia root*, A. rotunda, Eur.; and *Long Kunis wood*, a dye of an Aristolochia. Labuan.

Ord. Euphorbiaceæ. Spurge Family. (*Cases* 78, 79, *and* 80.)—Acrid and poisonous in an eminent degree, abounding in milky juice, which often yields caoutchouc.—*Euphorbium;* a gum resin, from Euphorbia officinarum (Africa), and E. Canariensis.—*Milk Hedge.* Inspissated juice of Euphorbia Tirucalli. Madras.—*Kattamandu gum*, of Euphorbia Kattamandu. Madras.—*Gums* and *resins* of other species of the genus.—*Sand-Box* tree; *fruit* and *flowers.* Hura crepitans. Of the immature capsules sand-boxes are made.—*Chinese Tallow.* Stillingia sebifera. China. *Seeds;* and *candles* made from the tallow.—Best *Caoutchouc* or *India-rubber*, from Pará, Siphonia elastica. Three cases are entirely devoted to the various preparations from this important substance, soft and vulcanized, presented by Messrs. Hancock and Macintosh. Imports of raw caoutchouc in 1853, 17,326 cwt., and, manufactured, from 4th June only, 629,499 lb.—*Candle tree; Berries* and *oil.* Aleurites triloba. Society Islands. The seeds yield a gum, which the natives chew. *Bracelets* made of them. A. laccifera is said to yield the gum lac of Ceylon.—*Physic-nuts.* Jatropha Curcas. S. Am.—*Cassava roots*, and various preparations, such as *Tapioca, common* and *pearl, Cassava Bread, Cassareep*, &c.; and *strainers* (long bags made of strips of palm-leaves), used for removing the poisonous juice from the pulp of the root. Thus a most abundant and wholesome food is derived from a very poisonous plant, Jatropha Manihot, by the aborigines of S. America.—*Castor-oil* and *seeds*. Ricinus communis. Its usefulness, combined with its palmate leaves, have given rise to the name, "*Palma Christi.*" 23,597 cwt. of the oil were imported in 1853.—*Wurrus*, or *Warus*. Röttlera tinctoria, imported from Bombay, as Dragon's blood.—*Seeds* and *oil* of Croton Tiglium.—*Cascarilla bark;* aromatic and powerful bitter. Croton Eleuteria. W. Ind.—*Box-wood;* ditto for wood-cutting, carved articles and *snuff-box* of the same. Buxus sempervirens. Eur. Imported in 1853, 3200 tons.—*African oak*, or *African teak;* neither an oak nor a teak, but a valuable naval timber, of this Family; it is Oldfieldia Africana. Trop. W. Afr.—*Wood* of Givodia röttleriformis, Madras; with various *figures* carved from it, and for which purpose this soft wood is well suited. (These figures represent various Indian trades, &c.)—*Jumping* or *moving seeds;* the lobes of a capsule of some Euphorbiaceous plant, from the Pacific side of America, which, in 1854, were first sent to England viâ Mexico, and excited a good deal of interest by their moving (by jerks) and almost jumping property. This was found afterwards to be occasioned by the sudden and peristaltic movements of the larva of a moth within, of which the egg must have been deposited in the state of the flower, for the shell had no perceptible aperture or wound whatever.—*Buah Bambi*. Fruit of Pierardia sativa. Java.—*Buah Cheremayi*. Fruit of Cicca disticha. Java.—*Buri nuts*. Hiptage Javanica. Amboyna.

Ord. Datiscaceæ. Datisca Family. (*Case* 80.) Said to be bitter and purgative.—*Akalbir* or *Ukalbere bark* of Kunawar is from Datisca cannabina. A yellow dye.

Ord. Nepenthaceæ. Pitcher-plant Family. (*Case* 80.)—These curious plants are described in the Garden Guide.—*Ascidia*, or *Pitchers*, of several species. Malay Islands.

Ord. Urticaceæ. Nettle Family. (*Cases* 80 to 85, including the sections or suborders.)—Worthless as Nettles are almost proverbially considered, and dreaded from the stinging property of many, especially the common European kinds, yet they yield a useful fibre, and a table vegetable, neither unwholesome nor unpleasant food. In the more enlarged sense, besides including, 1. *Urticeæ*, the true Nettles, the family contains also, 2. *Cannabineæ*, Hemp; 3. *Moreæ*, Mulberries and Figs; and 4. *Artocarpeæ*, the Bread-fruit. It is thus a very important Natural Order, including many esculent and useful plants. A large proportion abounds in textile fibres.—§ I. Urticeæ.—*Kentucky Hemp*. Urtica cannabina.—*Fibre of Neilgherry Nettle;* U. heterophylla.—*Fibre, thread,* and *lace*, made from the common *Stinging Nettle* in Ireland. U. dioica. — *Chinese grass*. The fibre of the Urtica nivea has received this name, and it is certainly among the most valuable of textiles for making calicos, &c., and is largely prepared by the Chinese.—*Rhea fibre* of India. Urtica Puya (the collection includes many other Urticaceous fibres).—*Wall-Pellitory, herb*. Parietaria officinalis. Eur.—§ II. CANNABINEÆ.— *Hemp*, from various European countries. Cannabis sativa, in temperate and northern regions is valuable chiefly for its fibre, and the consumption is enormous. In Africa and India it is scarcely known but as a powerful narcotic, and the natives indulge largely in it. Of the produce we possess: *Diamba*, and tincture of *Diamba*, from tropical Africa; *Gunjah*, and *Bang* or *Sidee*, from India; and *Hemp*, as smoked by the Caffirs, Natal. *German*, and *Russian*, and *Indian fibre* of *Hemp*; of which, from the two former countries, in 1853, were imported (rough) 940,022 cwt.; (dressed) 24,971 cwt.; and of Tow and Codilla of Hemp, 19,103 cwt.— *Hops*, and *extract* of Hops, and *Paper* made of Hop-bine (Humulus Lupulus) and of *Kentucky Hemp bagging*, U. S.; and *Hop-root*, from France.—§ III. MORACEÆ.—*Wood* and *fibre* of Morus alba, and wood of M. Tatarica.—*Tapé cloth*, of Broussonetia papyrifera, with instruments for beating the bark into cloth, and stamping or marking it. It constitutes the ordinary clothing of the S. Pacific islanders. — *Osage-Oranges*, the handsome, but not edible, fruit of Maclura aurantiaca. Arkansas. — *Fustic-wood*, and *chips*, Maclura tinctoria. W. Ind. —*Figs*. Ficus Carica. In their dried state, 50,244 cwt. were imported in 1853.— *Indian Caoutchouc, block* and *sheet*, and the *milk*, and *wood* from Ficus elastica; and *inspissated juice* from F. Indica. — *Caoutchouc*, and *skeleton leaves* of F. religiosa. India. —*Paper* made of a *Ficus* of India. — *Contrayerva Root* of Dorstenia Contrayerva. S. Am. — *Orange*, and *Liver*, and *stick*, and *seed*, and *white*, *thread*, *button*, and *block lacs*. These *lacs*, so valuable for dyes, are here placed among the Figs, because it is said "they are produced by the puncture of an insect called Coccus Lacca, or Coccus Ficus, upon the branches of several plants; as the Ficus religiosa, Ficus Indica, &c. &c., in various parts of India." The twigs become thickly encrusted with a mammillated resin, having a crystalline-looking fracture. Under the heads of *lac-dye*, and *shell-lac*, 46,562 cwt. were imported in 1853. — § IV. ARTOCARPEÆ. *Soft inner bark* of *Namaqua*, Olmedia armata, universally used for beds in the provinces of Choco and Darien. S. Am. — *Milk* of the *Cow-tree*. Galactodendron utile. Venezuela. The sap is used by the natives, and given to the children as cows' milk with us. — *Fruit* and *bark* of the *Upas* poison-tree, and concrete juice of the same; Antiaris toxicaria. Java. Inhabiting, as it does, the malaria regions of Java, this tree has a worse name than it deserves. —*Letter-wood* or *Leopard-wood,* Piratinera Guianensis. Central America.—*Natural sacks,* (*gunny bags*) of the *bark* of Lepurandra saccidora; Bombay; made by beating the fibrous bark, and peeling it off entire, leaving a section of the wood for the bottom; commonly used for holding rice. — *Fibrous bark* and *rope* of the *Trumpet Tree,* Cecropia peltata; and *Ambaubas,* or drums made of the hollowed portions of the trunk of the same, from San Gabriel; Rio Negro. Amazon. — *Bread-fruit;* and *fibrous bark* and *starch* and *biscuits* (Jamaica) — from the well-known Bread-fruit tree, Artocarpus incisa. — *Jack;* the gigantic fruit of Artocarpus integrifolia; the largest of eatable fruits, some have been known to weigh 80 lb.: the odour is said to be very disagreeable, the fruit good. *Kos gum* from the same. Ceylon.—*Tururi bark,* and an entire dress (*Tangas*) made of the same, Artocarpus. Rio Uaupés (Amazon). — *Shirts,* called *Túcaé,* made of two sorts of *Tururi bark;* one an Artocarpea, the other a Fig. Rio Uaupés (Amazon).

Ord. ULMACEÆ. ELM FAMILY. (*Case* 86.) —Chiefly useful for its *wood,* of which there are several kinds in the collection. *Elm bark.* Ulmus campestris; and *Slippery Elm bark,* of Ulmus fulva. U. S.—*Section* from the lower part of the trunk of the *Elm-Tree,* near which the *Duke of Wellington* stood at the battle of Waterloo, accompanied by the vouchers for its authenticity; together with a woodcut of the Tree immediately before it was felled and purchased by the late Mr. Children, of the British Museum.

Ord. BATIDEÆ. (*Case* 86.) —Batis maritima. Jamaica. Used in *pickle,* as Samphire in England.

Ord. PODOSTEMACEÆ. PODOSTEMON FAMILY. (*Case* 86.) — *Caaruru,* or salt from the burnt ashes of Podostemaceæ. Rio Uaupés. Brazil.

Ord. PIPERACEÆ. PEPPER FAMILY. (*Case* 86.) — Well known from its pungent property and its use as a condiment.—*Cubebs.* Cubeba officinalis. Imports, 48,511 lb. in 1853. They are aromatic, stimulant, and purgative.—*Black Pepper.* Piper nigrum. 5,496,885 lb. were imported in 1853. — *Long Pepper.* Piper longum. — *Kava; stems,* and *bowl* for making this intoxicating drink. This *pepper,* Piper methysticum, cut and chewed in the mouth, and the saliva ejected into the bowl* and rapidly fermented and strained, constitutes the favourite drink of the South Sea Islanders who call it "*Kava.*"—*Matico. Leaves,* closely reticulated, like a rasp, of Artanthe elongata. A powerful styptic. Peru.

Ord. AMENTACEÆ. CATKIN-BEARING FAMILY. (*Cases* 87–93.)

* A much larger bowl for the reception of this precious drink upon very grand occasions, being much too large to be admitted within the case, is placed near the adjacent wall of the room.

—Most extensive and important; chiefly composed of forest or timber trees, and possessing very varied properties: bitter, tonic, astringent, resinous, and balsamic; many yield eatable fruits or seeds. It is conveniently divided into groups or Suborders; but our space will only allow us to notice a few of each.—§ I. Betulaceæ. Birches and Alders: and these are chiefly valuable for their wood; common Birch for *rods* and *besoms*. Boxes made of Birch-bark. Paris. The barks are astringent, and used medicinally.—*Paper-Birch Bark.* Betula papyrifera, of N. Am., is used for various purposes. *Model* of a N. Am. *Canoe; Basket, mats,* and *card-case* made of it. *Cake* or *Bread* made of *Birch Bark* in Oregon.—*Soles* of *Shoes* made of Betula lenta. Sweden.—*Worsted* dyed with *Alder Bark,* Alnus glutinosa; and water-vessel made of the bark of Alnus incana. Kamtschatka.—§ II. Saliceæ. Willows and Poplars.—*Exchequer tallies* of Willow, accompanied by engraving and history.—*Willow Rods, peeled* and *unpeeled.* Chiefly used for *Basket-work,* of which several beautiful samples are in the Museum.—*Willow Bark.* Salix Russelliana. Eur. Astringent and tonic.—*Fishing-nets* and *cordage* made of Salix arctica by the Dog-Ribbed Indians of the Mackenzie River.—*Salicine.*—*Sussex Truck-baskets,* made at Hurstmonceaux, of Salix alba. *Toothpicks* of the same, from Madeira; and *clothes' pegs,* made in Suffolk, and sold at 1½*d.* the dozen!—"*Flat*" of *Willow-platting,* containing about eighty yards. *Willow shavings* for platting, dyed and undyed, and *baskets.*—*Tacamahac* of N. Am., the *resin* of Populus balsamifera. *Extract* of same.—*Boxes* made of wood of Populus Euphratica. Scinde.—§ III. Coryleæ, or Cupuliferæ. Nut and Oak Tribe. Under this head we have samples of *woods* of *Hornbeam, Beech, Oaks, Chestnuts;* and *Nuts, Mast,* and *Acorns* from several species of Corylus, Fagus, Castanea (Chestnuts), and Quercus. *Beech-mast oil* from Germany; and *Nut oil* of Corylus Avellana, France. *Boxes* made of Hasel-bark, Paris.— *Oak bark* for tanning, and *tannin,* and *bark* of *Cork-tree,* and samples of *Cork,* from Quercus Suber. Spain.— *Valonia,* "amata" and "camatina." Quercus Ægilops. For dyeing. Imports in 1853, 17,179 tons. S. Eur.—*Refuse Tan,* from *Oak bark.* Made into cakes for fuel in Brittany.—*Galls* of various kinds from *Oaks.* Among them the large *Mecca* or *Bussorah Galls,* called also *Apples of Sodom, Dead Sea apples:* used in the East (at Bussorah for example) for dyeing, and more esteemed than the common Nutgalls which are occasioned by the puncture of Cynips Gallæ tinctoriæ on the Quercus infectoria. The Mecca galls are formed by Cynips insana on the same oak. When on the tree, they are of a rich purple, and varnished over with a soft substance of the consistence of honey shining with a most brilliant lustre in the sun, which makes them appear like a delicious and tempting fruit. They are very astringent and bitter. The far-famed *Mad-apples,* Mala insana, or *Apples of Sodom,* Poma Sodomitia, of Josephus and other writers; "the fruit that never comes to ripeness" of the *Book of Wisdom,* and

"Which grew,
Near to that bituminous lake where Sodom flamed,"

and which, though beautiful to the eye, yet crumbles at the touch to dust and bitter ashes, was supposed by Hasselquist to be the Solanum Melongena (*Egg-plant*), by Seetzen to be a species of Cotton-tree, and by Mr. Lambert to be the galls here noticed.—*Gallic Acid.*—*Candles* made from a tallow or oil expressed from *Acorns* in N. Grenada.—§ IV. Myriceæ. Bog-Myrtle Family.—*Waxes.* The individuals of this group (species of Myrica) are remarkable for yielding a waxy substance in their berries, which is much used in commerce for candle-making, &c.—*Sweet-gale* leaves, Myrica Gale, and *Sweet-Fern leaves,* Comptonia aspleniifolia. Both from U. States. —§ V. Altingieæ.—*Balsam* of *Liquidambar.* Liquidambar styraciflua. Guatemala, and *oil* of the *bark* of Liquidambar Altingiana. Java.—*Storax-bark;* Liquidambar orientale. Asia minor. Used for fumigating.—§ VI. Plataneæ. Plane Trees.—*Wood* of *Button-wood,* of N. Am. Platanus occidentalis.

Ord. Casuarineæ. Beef-wood Family. (*Case* 93.) *She-oak-wood,* Casuarina quadrivalvis; and *He-wood,* C. stricta. Tasmania. Fine-grained Cabinet-woods.—*War clubs* and *pillows,* &c., made of Casuarina-woods of the Fejee Islands.

Ord. Juglandaceæ. Walnut and Hickory Family. (*Case* 93.) —Bitter and astringent, yet affording a valuable timber and esculent nuts; and these nuts yield largely a bland oil used for various purposes. —The collection contains samples of the woods, and fruits of various *Walnuts* and *Hickories,* and *Oil* of Juglans regia, &c.

Ord. Coniferæ. Cone-bearing, or Pine, Family. (*Cases* 93–100.)—An extensive arborescent family, yielding timber to a great extent, and turpentine, pitch, resin, &c.; chiefly inhabiting temperate climates. It is conveniently divided into three suborders: § I. Abietineæ. (Timber Trees, including the true Pines, Larch, Cedar, &c.) Our notices, where there is such a multitude of objects, must be brief.—"*Manna*" of Pinus excelsa, Simla; and *torches* made of the wood of the same, used in temples in Bootan.—*Common Turpentine, Pitch, Leaves* ("Pine needles," Germ.), "*Waldwool*" (fibre) and *cushion* stuffed with the same, and *soap,* and *extract,* from Bohemia; all of Pinus sylvestris. *Leaf* and *fibre* of "*Pitch Pine,*" of the Bahamas. Pinus Tæda.—*Pignons;* Seeds of *stone-pine,* Pinus Pinea, much eaten in Spain, Portugal, and Italy.—*Resin* of Pinus Khasyana. Khasya.—*American Turpentine,* from Pinus palustris. —*Oil of Turpentine,* and *Pine-oil* of various qualities; *Black, Yellow,* and *Amber resin; Stockholm* and *Archangel Tar, Spirits* of *Tar*, and *Tar acid*, and *Pine* and *black varnish* from various European Pines.—*Canada Balsam*, from Abies Balsamea.—*Esquimaux dish* of pine wood. Abies alba.—*Paper* made of the wood of the *Norway spruce*, Abies excelsa; and *Burgundy pitch* from same.—*Matches* in *block*, made of *Scotch pine*, Pinus sylvestris, used in Brittany. A vast consumption of this tree is caused by the trade in lucifer matches.*—*Kauri* or *Cowdie resin*, fine specimens, from Dammara australis, N. Zeal.; extensively purchased and used by the Americans.—"*Dammar resin*" of the Malays, from Dammara Australis.—*Resin* from Dammara macrophylla. Aneiteum.—*Pink* or *rose* and *white*, and *Guzzilam Dammar*, or resin, and "*Bdellium*" (Madras) supposed to be from species of Dammara (some probably from Dipterocarpeæ).—*Gum* or *resin* of Araucaria columnaris. Isle of Pines.—*Wood* of *Cedar* of *Lebanon*, Cedrus Libani (it has nothing to do with pencil cedar), and of *Deodar*, Cedrus Deodara, for comparison; and *Cedar Pitch* of the former. *Otto* or *attar* of *Cedar-wood*.—*Venice Turpentine* of Larix Europæa; and *Larch bark*, for tanning; and *netting* tanned with the same.—*Resin* of Microcachrys tetragona. Tasmania.—*Coniferous wood* dug up in the extreme arctic regions by Sir Edw. Belcher, R. N.—*Pine leaves* felted into compact balls in a lake in Shropshire, apparently by a peculiar action of the water (Lord Hill).—§ II. CUPRESSINEÆ. Cypresses, Junipers, &c., often shrubs.—*Juniper Berries* and *oil* of *Juniper*, much used in the manufacture of Hollands gin. Juniperus communis.—*Savin*; *wood*, and *powder*, and *oil* of same; from Juniperus Sabina. Eur.—*Gunisht resin*, from Juniperus phœnicea. Beloochistan.—*Pencil Cedar. Wood, raw*, and in different *states* of *preparation* to the perfect *pencil*, of Juniperus Bermudiana (frequently of J. Virginiana, or *red-Cedar*).—*Galls* on Juniperus Virginiana, U. S.—*Mat, shawl*, and *dress* made of "Cedar bark" of Oregon, Thuja gigantea.—*Sandarach*, a *Gum resin* from Callitris quadrivalvis; beautiful wood of the same. N. Afr.—*Wood* and *Cones* of Wellingtonia gigantea, and lithographic plates representing the most remarkable of those Mammoth trees, as they are seen in the native forest in the interior of California: their height varying from 300 to 400 feet, and the diameter of the stem from 30 to 36 feet. The bark is 22 inches thick. § III. TAXEÆ. Yews. Chiefly useful for the *wood*, which is very handsome. The common *Yew* is poisonous

* "The penny box of lucifers (or Congreves) is a real triumph of science, and an advance in civilisation. The first chemical light-producer, and a very complex affair too, cost a guinea! The best Norway deals are now, upon a large scale, cut into *splints* by machinery, and sold to the match-makers. The bundles are first dipt in sulphur at both ends, then in a preparation of phosphorus, or of chlorate of potash. Then the bunches are cut in two, and packed in deal boxes, made of splint-board, for sale. Two hundred thousand matches are cut, and 2000 boxes filled in a day, by one boy."—See *Dickens's Household Words*.

to cattle.—*Huon Pine-wood;* Dacnydium Franklinii, with various articles made therefrom, viz. a large and very handsome Cabinet on the first floor; filled with Rhizanthaceous plants.—In addition to the above and many other useful products of the Coniferæ, we can only mention the magnificent collection of *Pine-cones,* named, as far as can be, correctly; these are of great importance to nurserymen and planters, who compare them with cones they receive from abroad, and thus ascertain their proper names. The specimens of the cones of Araucarias and Dammars of the southern hemisphere are particularly valuable.

Ord. Cycadaceæ. Cycas Family. (*Case* 101.)—A very remarkable and most distinct Family, with somewhat the habit of Palms; bearing flowers and fruits in huge cones, sometimes 2—3 feet long; they yield starchy matter and mucilage. The heart or core of the crowns constitutes "*Caffer bread.*" Natives of the warm parts of Asia, America, and S. Africa, but frequently found fossilised in England.—*Cones* and *Fruit* (*Madu nuts* of Ceylon) and *Sago,* from the Cycas circinalis.—*Cones* and *models* of other Cycadaceæ.—*Zooloo* snuff-boxes of *seeds* of Zamia.—*Cone* of Stangeria paradoxa. Natal.

Ord. Rhizantheæ. Rhizanth Family. (*Case* 101.)—Of great interest in point of botanical structure. The collection is very considerable, and mostly placed in the Huon-Pine Cabinet on the first floor; and we must content ourselves with observing, that amongst them are a *drawing* and *model* (in a separate Case on the ground floor), both natural size, and *sections,* in liquid, of the gigantic blossom of Rafflesia Arnoldii, the largest known flower in the world: its diameter 3½ feet, and its weight 14lb.! Sumatra.—*Candles* made of the resin of a Balanophorous plant, from N. Grenada.

OLD MUSEUM BUILDING, No. 2.

CONTAINING

Cl. II. MONOCOTYLEDONS or ENDOGENS,

AND

Cl. III. CRYPTOGAMIC PLANTS.

First Floor. Room No. 1. (including *Cases* 1–9.)

Ord. Dioscoreaceæ. Yam Family. (*Case* 1.)—*Yam,* or *Igname; Model* and *Yam flour.* Roots of Dioscorea sativa.—*Affoo Yam starch,* from Dioscorea aculeata, Jamaica. Many species and varieties are cultivated in tropical climates as very nourishing and

PLAN OF OLD MUSEUM BUILDING.

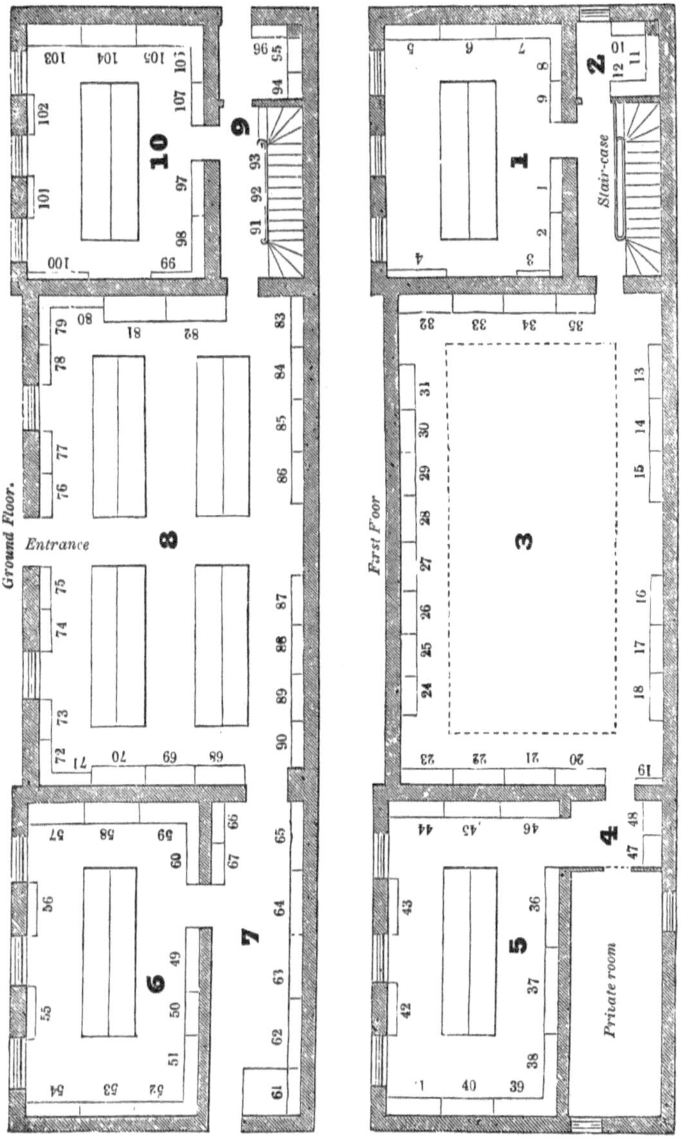

wholesome food. (The Japanese Yam, as it is called, recently offered at very high prices as a substitute for potato, is not likely to become a favourite food, nor is it suited to our climate.)—*Black Bryony roots.* Tamus communis. Eur.

Ord. SMILACEÆ. SARSAPARILLA FAMILY. (*Cases* 1, 2.) — Mucilaginous, bitter, and slightly acrid.—*Brazilian Sarsaparilla,* Smilax papyracea, in compact rolls (*fig.* 18.), bound with the curious pliable roots of an Aroideous plant called Oambé-címa (Spruce).—*China-root,* true, Smilax China, from China and Japan; S. *Pseudo-China* from Jamaica. Other kinds are: *Paraguay, Honduras, Guatemala* and *New Grenada, Laguayra, Natal,* &c. *Sarsaparillas.* Imports in 1853, 334,857 lb.

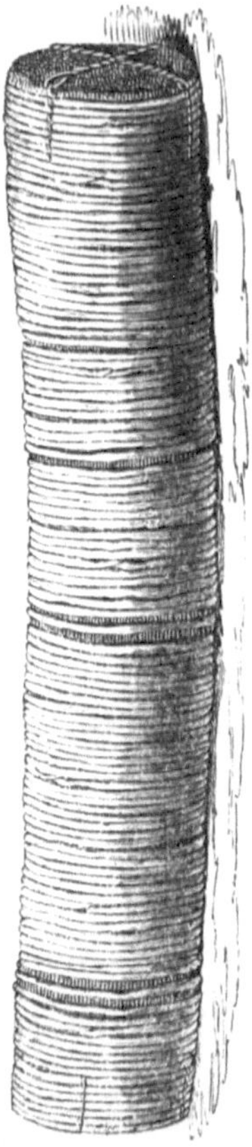

Fig. 18. SMILAX PAPYRACEA.

Ord. HYDROCHARIDEÆ. FROG-BIT FAMILY. (*Case* 2.) — *Anacharis Alsinastrum.* A small Pondweed-like plant, accidentally imported from Canada, where it appears to do no injury; while with us it has increased so much in our quiet rivers and canals as to impede navigation to an alarming extent, choking up the waters in Cambridgeshire, Lincolnshire, &c.

Ord. ORCHIDACEÆ. ORCHIS FAMILY. (*Cases* 2, 3.)—Yields but little to economic Botany.—*Vanilla.* The fragrant pods of Vanilla aromatica. Both Indies.—*Salep.* Tubers of Eulophia. India. In Europe *Salep* is from the roots of Orchis mascula, &c.—*Valerian* or *Mocassin root.* Cypripedium pubescens. N. Am.

Ord. ZINGIBERACEÆ. GINGER FAMILY. (*Cases* 3, 4.)—Herbaceous plants with fragrant foliage and tuberous or creeping roots or rhizomes, abounding in fecula. Aromatic and stimulant.—*Ginger,* Zingiber officinale; from various tropical countries; *preserved Ginger* and *Ginger starch.* Imports; *Raw* Ginger in 1853, 22,139 cwt., and *preserved,* 174,214 lb. — *Round Zedoary root,* Curcuma Zedoaria, and *Long Zedoary,* C. Zerumbet. India. — *Hill Arrow-root.* Curcuma montana. India. — *Turmeric root,* Curcuma longa, and *China round Turmeric,* Curc. rotunda (*fig.* 19.); and *Worsted* coloured with same. —*Turmeric powder,* made of Turmeric; Ginger, Cardamoms, Cayenne, Cloves, &c. Java.—*Attare* or *Atarei* (sometimes called Grain of Paradise. and Melegueta Pepper); *Fruit and Leaves.* Amomum

CURCUMA ROTUNDA.

Melegueta. Sierra Leone.—*Grain of Paradise.* Amomum Granum Paradisi.—Flowers and fruits of Amomum Danielli; together with A. cereum, and A. longiscapum, and A. latifolium. Sierra Leone.—*Korarima, Kheil* (Arabic). Fruits of Amomum Korarima. Soumali. —*Large round China Cardamom.* A. globosum?—*Hairy round China Cardamom.* A. villosum?.—*Xanthoid Cardamom,* A. xanthoides, and *bitter-seeded Cardamom.* China.—*Java Cardamoms.* Amomum maximum.—*Ceylon do.* Elettaria major.—*Malabar do.* E. Cardamomum.—*Ovoid China Cardamoms.* Alpinia alba.—*Galangal root.* Alpinia Galanga, &c. —*Flowers* of Costus Pisonis. W. Africa.

Ord. MARANTACEÆ. ARROW-ROOT FAMILY. (*Case* 5.)—Root or rhizomes amylaceous.—*Arrow-root,* from both Indies. Maranta arundinacea.—*Tous les mois; Starch,* or *Arrow-root* from Canna Indica, and other species of Indian-Shot.—Fine *matting* made of strips of the flower-stalks of Phrynium dichotomum. Sylhet.—*Baskets, head-dresses,* &c., of strips of *Uaruma,* the flower-stalks of a poisonous Maranta.

Ord. MUSACEÆ. PLANTAIN FAMILY. (*Cases* 6, 7.)—Yielding in some species valuable fruit, and probably in all abundant fibre, to which public attention is happily now directed.—*Banana* and *Plantain* fruits, vars. of Musa Paradisiaca and M. Sapientum.—*Sugar* of Plantains. A great variety of *Plantain fibre, cordage* and *textile materials, packing,* and *drawing* paper.—*Fibre* and two *costly shawls* and *flowering spike* of the *Manilla Hemp,* Musa textilis; called *Abaca* in the Philippine Islands, where it is a native, as well as at Mindanao, and also cultivated for its fibre; the fruit being worthless or very indifferent. From it are made the most beautiful shawls, &c.; and it is largely exported, especially to the United States, where the coarser kind is manufactured into "White Rope," chiefly at or near Boston, and the old rope again is made into excellent paper. The entire quantity of fibre raised at Manilla in 1840 was 83,790 piculs, and in 1841, 87,000, which sells at from 4 to 5 dollars the picul; and in those two years were sent to the U. States alone, 68,280 piculs; and in 1841, 62,700 piculs, valued at 200,000 dollars. Twenty thousand piculs only are said to go annually to Europe. (See Wilkes's Voyage of the U. S. Exploring Expedition.) —*Fruits* of the curious *Ensete* of Bruce (Musa Ensete). Abyssinia.

Ord. IRIDACEÆ. IRIS FAMILY. (*Case* 8.)—The roots or rhizomes of some are fragrant and stimulant. — *Orris root.* Iris Florentina. Much used in perfumery, &c.—*Saffron.* The dried *stigmas* of Crocus sativus, of which 83,453 lb. were imported in 1853. Contains an active volatile oil, and is used medicinally and as a colour. In the loose state it is called "*Hay Saffron;*" in mass "*Cake Saffron.*"—*Root*, medicinal, of Homeria collina. Cape.

Ord. Amaryllidaceæ. Amaryllis Family. (*Case* 8.) — Narcotic poisons are prevalent in many of this family. — *American Aloe fibre.* Agave Americana. We have this from various localities, and much that comes under the name of "Aloe fibre," and various articles of clothing, which we suspect to belong to this; other *fibres* of Agave Mexicana, and A. diacantha, and vivipara (called *Silk-grass fibre*). *Razor strops,* made of portions of the stem of the latter, used in Brazil. — *Fibre* called *Seemay Katholay,* from Fourcroya gigantea. Madras.

Ord. Bromeliaceæ. Pine-Apple Family. (*Cases* 8, 9.)—One species of this family yields the well-known Pine-apple, and all perhaps afford *fibre,* of which there are various kinds, and *cloth* and *twine* from several species, in the collection. — *Tillandsia usneoïdes* from the warm parts of N. and S. America; extensively used for stuffing mattresses, &c.

First Floor. Rooms No. 2. and 3.

Ord. Gramineæ. Grass Family. (*Cases* 10–35.) — One of the most extensive and useful of all the Natural Orders, yielding food for man and fodder for cattle, in almost every region of the globe. Rooms No. 2. and 3. (Gallery) are wholly devoted to its products, most of which formed part of the great Agricultural Collection of Messrs. Lawson, to whose volume we must refer for *particulars* of all that are cultivated in Europe, viz. the numerous *Wheats, Barleys, Rye, Oats;* in short, all the *Cerealia,* and their products, including *Macaroni* and *Vermicelli* (of which 5334 cwt. were imported in 1853), and all the *Grasses* reared for *herbivorous animals.*—Our own additions are too abundant, even to enumerate in this Catalogue. Some of the more interesting may be specified.—A small room (No. 2. *Cases* 10–12) is entirely filled with the products and articles derived from *Bamboo,* Bambusa arundinacea, among them *Leaves of Bamboo,* used by the Chinese for the inner lining of their chests of Tea; different *varieties of canes; baskets, ornaments,* and *images,* of the rhizome or rootstock; Chinese *chop-sticks, fans, flutes,* and other musical instruments; Chinese *shirt* of split Bamboo; *chairs, umbrellas, Tabasheer,* mineral concretions found in the interior of the living stems, &c.

Room No. 3. (the Gallery) is occupied by the rest of the Grasses (*Cases* 13–35.), including—*Platting* from the stems of numerous kinds; of the *Marram* grass of our coasts, Ammophila arenaria; of the *Esparto*, Lygeum Spartum, of Spain and Portugal; of Rye and other *straw*, made in Orkney, St. Alban's, Nice, &c.—A fine collection of *Maize* or *Indian Corn* (some from the Peruvian tombs of the Incas), *Millets* (some *in rolls* from Baikie's Niger Expedition). —*Farina and bread* made of the core or heart of the ears of Maize. Austria. — *Rice*, &c. — Models of apparatus for making *Malt* (of which the imports in 1853, were 1,504,629 cwt. clean, and 19,477 qr. rough); and the proportion of *malt* and *hops* employed by Hanbury and Co., in brewing porter. — *Sugars**, with samples and models of the Sugar-cane. ("To what purpose cometh there to me incense from Sheba, and the *sweet cane* from a far country." —*Jeremiah*, vi. 20.) *Specimens* illustrative of making paper from same, and various *Cane Papers*.—*Starches* from the various Cerealia. — *Flour*, that had been buried by Capt. Beechey R.N., in 1824, at Chamisso Island, Kotzebue's Sound, for the use of Sir John Franklin; and when dug up by Capt. Kellett, in 1849, it proved perfectly sound.—*Macaroni* and *Vermicelli* from the finest flour of wheat. — *Darnel* (Lolium temulentum), said to be poisonous, and bread made from it. — *Slag* formed in the furnaces from the use of waste sugar-cane as fuel.—*Brushes* commonly made in Venice and Trieste of the panicles, called *whisks*, of Sorghum. *Sugar* from Sorghum saccharatum, Natal, where attempts have been made to render it profitable. — *Oil* of *Citronella*, made from Andropogon Schœnanthus, and other Indian species. — *Cuscus Grass* of India, much used on account of its fragrance. Andropogon muricatus. — *Canada-Rice*. Zizania aquatica. — *Paper* made of various grasses, and *Indian Arrows* of same.—*Ergot* of wheat and of other grasses.—*Rosaries*, &c., of *Job's tears*, seeds of Coix Lacryma. India.—*Wheat* and *Barley* from Egyptian tombs, &c.— *Bird's-nest* made of *Guinea-grass*, Panicum maximum.

First Floor. Room No. 5.

Ord. Liliaceæ. Lily Family. (*Cases* 36–38. and also 66, 67. of Room 7.)—An extensive family, divided by some into many groups or Suborders, by not very natural characters; beautiful generally in their blossoms; many yield valuable fibres, or resinous matter; not a few are purgatives, or stimulants, or emetics, or diaphoretics, and employed medicinally. The Tubers are often

* During 1854 the total quantity of brown Muscovado sugar entered for home consumption, in the United Kingdom, amounted to 2,176,736 cwt. (at a duty of 11*s*. per cwt.) Of this sugar 901,857 cwt. were imported from the British W. Indies; 238,723 cwt. from the Mauritius; 382,709 from Cuba; 307,365 cwt. from the Brazils; 68,285 cwt. from the E. Indies; and 129,111 cwt. from the Philippine Islands. The total quantity of yellow Muscovado and brown clayed sugar entered for home consumption, amounted to 2,157,104 cwt.; of which 749,457 cwt. came from Cuba; 244,303 cwt. from Porto Rico; 35,559 cwt. from Brazil; 120,450 from the Mauritius; 194,137 cwt. from the E. Indies; and 703,280 cwt. from the British W. Indies.

farinaceous and esculent. — *North-west Rice* (and *bread* made of it); this name is given in N.W. America to the roots of Fritillaria Kamtschatkensis. — *N. Zealand Flax.* Phormium tenax. An extensive collection of the fibres and manufactured articles of this Flax, part made by the aborigines, and part by European settlers. — *Bow-string Hemp. Fibre* of two species of Sanseviera, S. Guineensis, Trop. Africa; and S. Zeylanica, India.—My attention has recently been directed by the Foreign Office to *fibre, cordage,* and *string* lately received from Loando, W. Africa, together with living plants, called "*Iffé,*" of a new Sanseviera (S. Angolensis); the leaves are three or more feet long, an inch and more thick in the lower portion, terete and striated; abounding in strong fibre. — *Fibres* from Madras, said to be from *Aloe;* for the names of A. littoralis, and A. spicata are given; from Accra, Africa, and from Arabia; but whether derived from true Aloes, or Aloe-like plants, is doubtful. — *Aloes* (*inspissated juice*), a powerful cathartic, from Aloe Barbadensis, A. purpurascens, A. Soccotrina, and A. ferox; together with *liquid Aloes,* and *Aloin.* Imports of Aloes in 1853, 246,575 lb. — *Fibres,* from Yucca angustifolia, Y. aloifolia and Y. gloriosa. — *Garlics* and *Onions. Seed* and *Drawings* (in Room No. 7.) of different species and varieties.—*Squill, root* (*bulb*). Urginea Scilla. — *Tchirish; root* and *powder.* Eremurus spectabilis? The roots afford a valuable gluten much used in Turkey. Erzeroum. — *Black-boys. Stems* and *gum-resin,* and *flowering heads* of species of Xanthorrhœa, the *Grass-gum trees* of Australia (in *Cases* 66, 67. Room No. 7.).—*Garment* of the leaves of Cordyline utilis, dyed black with the bark of a Fagus, from N. Zealand. — *Dragon's blood* of Dracæna Draco, from a garden at Lisbon, and also from the celebrated tree of Orotava. (See section of branch and drawing of the latter in passage at the top of the staircase.)—*Dress* made of the *leaves* of Dracæna terminalis. Navigator's Islands.—*Solomon's seal; root* and *fibre;* and *lace* of the stem of Polygonatum multiflorum, prepared in Ireland.

Ord. Melanthaceæ. Colchicum Family. (*Case* 38.)—Acrid, purgative, emetic, and narcotic, sometimes poisonous. — *White Hellebore root,* and *Veratrine* derived from it. Veratrum album. Eur. —*Cevadilla* or *Sabadilla; fruit* and *seeds* of Asagræa officinalis; much used in neuralgia and rheumatism. *Hermodactyl roots.* — *Colchicum; Roots* or *powder* of Colchicum autumnale (*fig.* 20.), a narcotic acrid poison, much used medicinally for gout, &c. Eur.

Ord. Zosteraceæ. Sea-wrack Family. (*Case* 39.)—"*Alva*" (Ulva) *marina* or *Grass-wrack.* Zostera marina. Collected on sea-shores and salt-marshes, and much used, especially in N. Europe, Iceland, &c., as stuffing for beds and mattresses.

Ord. PANDANACEÆ. SCREW-PINE FAMILY. (*Cases* 39, 40.)—Many of these remarkable plants are natives of muddy shores of tropical rivers, and have aerial roots, which descend like buttresses, and prevent their being washed away by the currents. The leaves, as may be here seen, are manufactured into *ropes, hats,* &c.—The *Sugar-bags* of *Mauritius* are made of the leaves of Pandanus utilis.—*Oil* from flowers of Pandanus inermis. Amboyna. — *Panama hats,* made of leaves of Carludovica palmata. Cigar-cases, very fine and beautiful, of the same.

Fig. 20.

COLCHICUM AUTUMNALE.

Ord. JUNCACEÆ. RUSH FAMILY. (*Case* 41.)—Of the common *Rushes* (Juncus) a great trade is carried on in certain counties in England, in the preparation of the pliable stems for baskets, besoms, chair-bottoms, mats, &c., and the working in those materials, and no less for making candles, *rush-lights,* of the *pith,* in rural districts; chiefly of Juncus effusus; and at Norwich there is an annual chartered Fair held, where about 800*l.* is expended in buying the same article from the peasantry. The *raw material,* as there sold, and the prepared *rush* for candles, are in the collection; and a most beautiful specimen of artistic work of a Chinese temple, made and presented by Mrs. Kingdon, of Exeter, is entirely of this material.—*Palmite, leaves* and *stem,* with *hats, baskets, broom, brushes,* &c., made of the strong fibre. Prionium Palmita. S. Africa.

Ord. ARACEÆ. ARUM FAMILY.* (*Cases* 42. *in part, and* 44.)—Acrid and poisonous; yet when cooked, both the leaves and farinaceous tubers are common articles of food in warm countries.—*Portland Arrow-root* and *rhizomes* of Arum maculatum or *Wake Robin;* formerly rather extensively prepared and sold at Portland Island.—*Starch* from the rhizomes of Colocasia antiquorum. A common article of food in the W. Indies, Africa, Pacific Islands, &c., where it is known as *Taro.* — *White Coco Biscuits,* made of Colocasia esculenta. Jamaica. —*Rhizome* and *stem* and *foliage* of "*Gladden,*" or *Reed-mace.* Typha angustifolia. The latter (leaves) much used in Norfolk for mats, basket-work, &c.—*Basket* from Scinde made of Typha elephantina † (T. angustifolia ?); and *cakes* made of the *pollen* of the same, sold in the

* Including TYPHACEÆ.

† This is possibly the "bulrush" of Scripture, of which the basket was made for the infant Moses, such being still in common use in Eastern countries. Some authors consider the Papyrus to have been the bulrush of Holy Writ.

bazaars of Scinde as food!—*Cakes* of the *pollen* of Typha angustifolia; made and eaten by the aborigines of N. Zealand! That edible cakes should be made of such a material, is a remarkable circumstance; still more remarkable that people so remote from each other, and so dissimilar in habits and manners as the Scindians and N. Zealanders, should use the same article, and for the same purpose. The cakes are identically the same; yellow, almost like gamboge.—"*Black weed*" so called in Norfolk, *leaves* of Sparganum simplex; used for binding the edges of matting, baskets, &c.—*Oil* of the fragrant *Sweet-Sedge*, Acorus Calamus. Eur.

Ord. CYPERACEÆ. SEDGE FAMILY. (*Cases* 43, 45, *and* 46.)—No medicinal virtues, but the pliable culms or stems and leaves are excellent for weaving into mats, hats, &c., and one kind yielded the paper of the ancients called *Papyrus*, or *Egyptian Papyrus*, which, however, as well as the famed Egyptian Lotus flower, has wholly disappeared from the rivers and marshes of Egypt! "How literally," says Dr. Bromfield ("Letters from the East," ined.), "have the words of the prophet been fulfilled in this single and apparently unimportant, as in so many more remarkable and weighty instances,—the reeds and flags shall wither, the *paper reeds* by the brooks, by the mouth of the brooks, and every thing sown by the brooks shall wither, be driven away, and *be no more*." * The pithy substance of Cyperus Papyrus formed this paper. Our specimens are from the Tombs at Thebes (Rev. C. C. Babington); and *modern Papyrus*, made from the Sicilian Papyrus, Cyperus Syriacus (Prof. Parlatore). — *Tubers* of Cyperus esculentus. S. Am.—"*Boulder*" (Norfolk), or *Bull-rush;* Scirpus lacustris, as sold in bundles for making *Chair-bottoms, Horse-collars, Bee-hives, Mats, Hassocks, Baskets,* &c. *Great-coat*, from Portugal, supposed to be of the same material.—*Hassocks* of the natural tufts of Carex paniculata. *Mats* and *brooms of the leaves* of the same.

GROUND FLOOR. ROOMS NO. 6, 7, AND 8.

Ord. PALMACEÆ. PALM FAMILY. (*Cases* 49–65, *and* 68–90.)—It would require a volume to notice all the benefits that mankind derives from Palms.

> "The *Indian nut* alone
> Is clothing, meat and trencher, drink and pan,
> Boat, cable, sail, and needle, all in one."

They yield timber, fibre of every variety, oil, wax, starch, sugar, daily food, mild and intoxicating drink: it is rather difficult to say what they do not afford, and impossible in a brief catalogue to notice a tithe of the contents of this collection. We

* Isaiah xix. 6, 7. See also Bromf. Letters, p. 181.

can only direct attention to a few of the remarkable kinds.—§ I. ARECINEÆ.—*Combs* from the Uaupés river (Amazon), made of the wood of the *Bacaba Palm*, Œnocarpus Bacaba, bound with strips of Gynerium saccharoides. — *Betel-nuts.* Areca Catechu, and *charcoal* from the same, and *walking-sticks* of the *seeds* (strung on an axis), and *leaf-stalk* used for writing on in India.—*Spatha* and *baskets* of the *leaves* of the *Cabbage Palm.* Areca montana. Jamaica. — *Spadix*, and *fruit*, and *musical instruments* made of the stem of the *Paxiube Palm.* Iriartea exorrhiza: together with *male* and *female spadices* and *aerial roots*. Brazils. —*Stem* and *wax*, and *candles* prepared from Humboldt's *Wax Palm.* Ceroxylon Andicola. N. Grenada. —*Ejoo* or *Gomuti fibre* of Saguerus saccharifera, and *Saguero*, a palm wine, from the same. Malacca.—*Gum* of Caryota horrida. Ceylon. — § II. CALAMEÆ.—*Manilla, Malacca,* and *Rattan canes*, much used for chair-bottoms, and light furniture from different species of Calamus. *Pillow*, made of the same, and *cable* of a Chinese junk. Malacca. —*Spadices* of Plectocomia elongata. Java.—*Sago; Flour* and *starch*, and *bread*, from Sagus lævis; Singapore, &c.; of which 94,703 cwt. were imported in 1853.—*Mats* made of S. Ruffia. Samples of *Sago* in five different stages of preparation from Sagus Rumphii. Amboyna.—*Spadix* and *fruit* and *fibre* of Mauritia flexuosa. — *Wine Palm* of Pará. Mauritia vinifera.—§ III. BORASSINEÆ. — *Palmyra Palm.* Borassus flabelliformis; *fruit* and *leaves*; and *Fans, books*, &c., made of the leaves. A *Burmese Ms.*, believed to be of considerable antiquity, made of the plants or strips of the *Palmyra*, written on with a stylus, and done up as a book, or rather fan-fashioned, the edges painted and gilt. Presented by W. Borrer, Esq. India.— *Double* or *Maldive Cocoa-nuts.* Lodoicea Seychellarum. *Spadices* of the flowers, and nuts, wood, &c. *Baskets* of the leaves. Seychelles. Long supposed to be a submarine palm; and then the fruits had all kinds of properties attributed to them, such as being antidotes to poison, &c., and they sold at enormous prices.

> "Wide forests there beneath Maldivia's tide
> From withering air their wondrous foliage hide.
> The green-hair'd Nereids tend the bowery dells,
> Whose wondrous fruitage poisonous rage expels." *Lusiad.*

— *Doum Palm. Fruit*, edible, having the flavour of gingerbread, and *forked stem* of Hyphæne Thebaica. Upper Egypt. — *Spatha* and *spadix*, and *caps* made of the *spatha* of Manicaria saccifera. B. Guiana.— § IV. CORYPHINEÆ. *Covering* of a *tent* made of the *Talipot* leaf. Corypha umbraculifera. Ceylon.—"Penang *Lawyers*," the *Canes*, of Licuala acutifida.—*Caranauba palm.* Copernicia cerifera. Curiously embossed *stem* with its covering of *wax*; and collected *wax*; *leaves* and *mat*, and *hats* made therefrom. Brazil.— *Baskets, hats*, &c., of the *leaves* of Chamærops humilis. Sicily.—*Leaves* and *Chinese Great coat*, and *hat* of the *fibre* of Chamærops excelsa; China; and *brush* of the same.—*Baskets, hats, cigar-cases, &c.*, of leaves of Chamærops argentea of Cuba; made at St. Albans.—*Ground Rattans*, and *spadix* of Rhaphis flabelliformis.—*Dates* of various kinds, and *stem*, and *walking-stick* of the midrib of the leaf of Phœnix dactylifera; and *date cakes* made by the Arabs. Syria.—Beautiful *Kaffir baskets* from Natal, of leaves of Phœnix reclinata.—§ V. Cocoineæ.—In this group are *flowers* and *spathas* and *fruit* and *foliage* of species of Bactris; of Gulielma, or the *Peach Palm;* and of Acrocomia and Astrocaryum; and *fibre*, and *baskets*, and *hats*, and *various ornamental work*, especially of the latter genus; all from S. Am.—*Stem of a Munbak Palm*, Astrocaryum Munbacca, encased by the natural growth around it of a fig.—*Piacaba fibre*, and *nuts* and *ornaments* from the *Coquilla Palm*, and *brooms* and *brushes*. Attalea funifera. Brazil. Of these nuts there is a great variety of articles made; handles for bell-pulls, &c.; and of the fibre the consumption in Europe is now enormous. But similar fibre of another palm (Leopoldinia Piassaba) goes under the same name.—*Box* made of the leaves of *Uanassu Palm*, Attalea sp. Rio Uaupés.—*Cohune nuts* and *oil*, of Attalea Cohune. Honduras.—*African Oil Palm*. Elais Guineensis. Of this valuable palm, most important for its oleaginous product, we have a fine series of specimens, exhibiting the entire *cluster* of *fruits*, *dry* and in *liquid*, dissections of the *seeds*, and various preparations from the *raw oil*, as imported (636,628 cwt. in 1853) from the African coast, to the perfect candle and night-light, together with beautiful specimens of *palmetic acid* and *glycerine*, all presented by Messrs. Price and Co.—*Spadix* and *fibre*, and *band*, and "*noli*" (or scurf of the petiole, used for tinder) of Elais melanococca of Brazil, which Mr. Spruce says is exceedingly like the E. Guineensis and would be found probably to yield an equally valuable oil.—*Cocoa-nuts*. Cocos nucifera. The most celebrated and best-known of the Palms; for some of the many uses to which the different parts of the tree are appropriated, we must refer to the named articles.—*Implements* used in collecting the *toddy* from Palms, *models*, *drawing*, &c. Of the oil alone, 164,196 cwt. were imported in 1853, and of the fibre, as *coir*, 50,665 cwt. The two following are of dubious arrangement among the Palms, viz. *Male spadices*, and the fine handsome *fruit* of Nipa fruticans: and *Ivory-nut Palm*, Phytelephas macrocarpa, from Magdalena, its *fruits*, *seeds*, *spatha*, and *leaf;* and articles made by the Indians, and others made in Europe of the ivory-like albumen.—Besides products of known *Palms*, we have a multitude of manufactured articles from species that are not yet known to us by name.

Cl. III. ACOTYLEDONS or CRYPTOGAMS.

Room No. 8.

(Great Room, Ground Floor, in Table Cases.)

As already observed, the products of this natural Class are neither so interesting nor so numerous as those of the two previous Classes. Nevertheless several claim attention.

Ord. Filices. Fern Family. — Some yield food; a few are employed medicinally, as demulcent and astringent. — *Male Fern-root*, and *oil* of same. Aspidium Filix-mas. Eur.— *Fern-root* of N. Zealand, where it much eaten. Pteris esculenta. — *Maiden-hair*, from which syrup of capillaire is prepared. Adiantum Capillus Veneris.— *Hygrometric Club-moss:* Lycopodium lepidophyllum. Remarkable for rolling up like a ball, in dry weather, and spreading widely open in wet. W. Mexico. — *Door-mat* made of *Club-moss*, Lycopodium clavatum. Sweden. — *Dutch Rushes;* used for scouring and polishing, on account of the quantity of siliceous matter in the epiderm. Equisetum hyemale. (*Fig.* 21.) Eur.

Fig. 21.

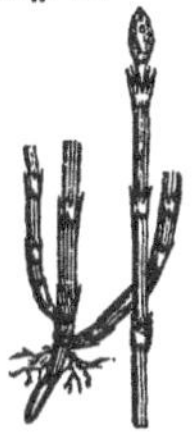

EQUISETUM HYEMALE.

Ord. Musci. Moss Family. No active properties.— *Bog-moss*, Sphagnum latifolium (*fig.* 22.), and other species, absorb and retain moisture: hence useful in packing living plants for long voyages; and much used for packing Lapp children, who are swathed in it till they are old enough to take care of themselves.— *Hassocks*, much used in churches in the North of England, of tufts of the *great Hair-moss*, Polytrichum commune, simply cut from a moss-bog: the plants held together by the matted roots. *Basket* woven and brooms made of the stems of the same. Yorkshire.— *Wicks* for burning in oil, used by the Esquimaux, made of slender tufts of Dicranum condensatum. — *Peat* from Norfolk, formed of Hypnum filicinum.

Fig. 22.

Fig. 23.

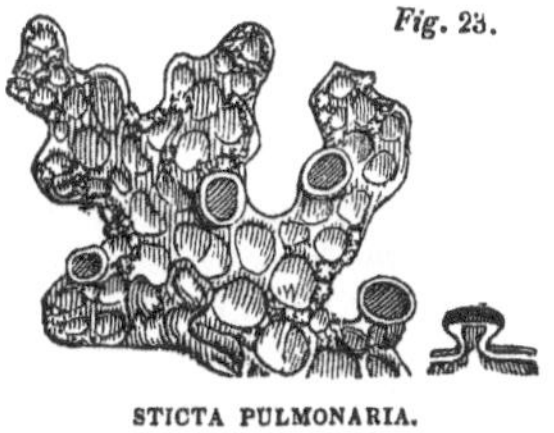

STICTA PULMONARIA.

Ord. Lichenes. Lichen Family. — Several Lichens yield a nutritious matter called Lichenin or Lichen-starch, often accompanied by a very powerful bitter. Many yield dyes; some are eaten.—*Lung-wort.* Sticta pulmonaria. (*Fig.* 23.) Eur. From a fancied resemblance to the lungs, it has been used in diseases of that organ:—and *Liverwort,* Peltidea canina, for hepatic diseases.— *Iceland-moss.* Cetraria Islandica. In Iceland used as food (the bitter principle extracted by steeping in water), in other countries as medicine.— *Rock-manna:* or, as some think, the miraculous Manna of the desert, Lecanora esculenta, from Persia and Arabia. — *Orchil-weed.* Roccella tinctoria (*fig.* 24.), from various countries; of this 16,909 cwt. were imported for dyeing purposes in 1853.—*Litmus,* and various *stuffs dyed* therewith. — *Cudbear,* and *stuffs* dyed with it. Lecanora tartarea.—*Perelle,* and *dyed stuffs.* Lecanora Perella. — *Reindeer-moss.* Cenomyce rangiferina: the chief food of the Reindeer in Arctic and Subarctic regions, where this Lichen had been supposed to be peculiar; but there are very fine specimens in the Museum from tropical America, and from Australia in the Southern hemisphere. — *Tripe de Roche* of the Canadian voyageurs, species of Gyrophora; a substitute for better food in Arctic America, and often the only sustenance of the travellers, as in Franklin's and Richardson's journeys.—*Lichen-bread,* used in Finland in times of scarcity.

Fig. 24.

ROCCELLA TINCTORIA.

Ord. FUNGI. MUSHROOM FAMILY.—Yields food and poison, and many are exceedingly curious in structure. Very many are parasites on dead or dying plants, wood, &c. — *Geastrums* (see also *drawing*) are very curious in their forms. — *Mushrooms. Drawings* and *specimens, dried* and in *powder,* and *spawn* in the earth giving the soil a whitish colour. Agaricus campestris, and A. Georgii.—*Truffles,* familiar to gourmands. Tuber cibarium. —*Fly-Agaric.* Amanita muscaria: so called because a decoction in milk is used to decoy and destroy flies; and the Tartars of Siberia use the plants for intoxication. — *Morelles.* Morchella esculenta.—"*Agaric,*" of the shops, a well-known styptic. Polyporus officinalis. — *Amadou,* or *German tinder,* prepared from Polyporus fomentarius.—*Fungus Caterpillars,* as they are called. Various species of Sphæria, parasitic on dead caterpillars. — *Giant Puff-ball.* Lycoperdon giganteum. These great balls have been known to be the growth of a single night, on heaths. *Ears* of *wheat,* and other Cereals, injured by Fungi, such as Smut, Brand, Bladder-brand, &c., and an excellent drawing made with the *Smut* (Uredo Segetum).—*Dry-rot* (it ought rather to be called the Wet-rot). Merulius lachrymans (*fig.* 25.); a destructive Fungus, encouraged by dampness and absence

Fig. 25.

MERULIUS LACHRYMANS.

of ventilation. — *Wood* (*oak*), stained green naturally by the *mycelium*, or spawn, of a small fungus, Peziza æruginosa (*fig.* 26.), and *spheres* of the wood, polished. Brought into use at the Tonbridge-ware manufactory. — And lastly, I have only space to notice the *Vinegar plant*, or *Mother of Vinegar*, here seen in a living state, floating upon a liquid of sugar and water. This is a minute fungus, allied to the *Mucor* or *Mould*, Penicillium glaucum, of which the mycelium or spawn forms a tough leathery web. A bit of this thrown into the above liquid rapidly increases, induces acetous fermentation, and changes the sugar and water into vinegar. (The Yeast plant, or "Mother Yeast," a substance not so easily preserved, is also considered a Penicillium; and to its action is due the formation of yeast.) — A very fine collection of Fungi presented by Sir James Brooke from Borneo.

Fig. 26.

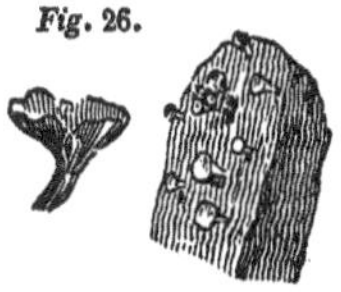

PEZIZA ÆRUGINOSA.

Ord. ALGÆ. SEA-WEED FAMILY. — Aquatics both of salt and fresh water. The least organised of this Family of Plants (Algæ) are most allied to the lowest in the scale of animal life, and the line of demarcation between them is not easily defined. Many are used as food, or for manure. They yield Gelatine; the marine species Iodine, Kelp, &c. — *Trumpet Sea-weed.* Ecklonia buccinalis. Cape. A *trumpet* formed of the hollow stem of this gigantic sea-weed, the most bulky of all Algæ.—*Fronds* and *bladders* of Macrocystis pyrifera. Antarctic seas.—Portions of *Durvillea utilis.* A single leaf or frond is more than a man can lift. In S. Chili this plant is used for food. This curious production has the longest stem of any known vegetable, from 500 to 1500 feet in length!—*Kelp, iodine, muriate* and *sulphate* of *potash, alkaline salt,* &c., from Fucus vesiculosus, and other species.—*Agar Agar* (Sphærococcus spinosus), *Ceylon Moss* (Plocaria candida), *Irish Moss* (Chondrus crispus), and other Algæ yielding gelatine, which is employed as food (blanc-mange, &c.), and more extensively for dressing silks and calicos. Some are of opinion that this kind of substance forms the edible swallows' nests of the Malay Islands; but *these* by the ablest Zoologists are considered to be entirely an animal product derived from the bird. —*Gulf-weed.* Sargassum bacciferum: floating with the gulf-stream, by means of its numerous little bladders, never known to be attached to any place of growth, and so abundant in some localities as to impede the sailing of ships.—*Globe Conferva,* Conferva ægagropila, a curious filamentous green fresh-water plant, whose growth is always in the form of a ball, as large as a small orange, floating on the surface of small lakes in Shropshire.—Some drawings on the walls are illustrative of the eatable *Dulse* and *Laver* (Ulva Lactuca); and of the *Red Snow* of Arctic regions, Protococcus nivalis; the simplest of all vegetable forms, a single plant being a minute solitary spherical cell, and flourishing in the midst of Arctic or alpine snows.

MISCELLANEOUS.

Under the manifestly unscientific head of "Miscellaneous" we are compelled to include a large collection of yet unnamed *Woods*, of British and foreign origin, many beautifully prepared, and more or less cut and polished. The other miscellaneous objects, which from various causes cannot conveniently be displayed in the arranged collections, we shall briefly notice as far as our space will allow; indicating the rooms where they are placed.

Room No. 1. (on the Table). *Artificial flowers.* Groups made of *different materials; wax, cambric, silk, shells,* &c., and more or less true to nature. Botany contributes its share to this art; and that it is not of trifling consequence may be seen by the following extract from Dickens' Household Words: — "The total manufacture of *Cambric* flowers in Paris alone, in 1847, was prodigious, amounting in value to more than 400,000*l.* sterling. We, in England, only took 12,300*l.** worth of this value. The cambric, muslin, gauze, velvet, silk, and other materials are procured from St. Etienne, St. Quentin, and Lyons; the dyes and colours are prepared expressly for the purpose by manufacturing chemists; the buds, leaves, petals, stamens, pistils, and other component parts, are made in small workshops by persons who attend to only one part of the flower, while the whole is fitted together in other workshops. Even these workshops are frequently limited to one single kind of flower each, so completely is the division of labour carried out. There are about fifty small manufacturers of petals and stamens, and other component parts, employing about 500 persons; while there are nearly 600 dealers or venders, who employ nearly 6000 persons in building up the various integers into whole groups of flowers. Of this immense number of persons, about 5000 are women, whose average earning is estimated at 1*s.* 8*d.* per day. Several of the manufacturers effect sales to the amount of 10,000*l.* a year each. We must therefore regard French flower-manufacturers as commercial men of notable import." — A Case of *Starches* (prepared from *Rice*), and *Indigo cakes.*—Ditto of *Essential Oils.*—Ditto of Ashby's *Groats, Barleys,* &c.—Box of *Spices.*—Case of *Infusions* and *Decoctions.*—Collection of prepared *Chocolates* and *Cocoa,* with *Vanilla.*—Case of *Indian Spices* and *Cur-*

* This amount may be so, from France: but certain it is that the returns of imported artificial flowers for 1853 given by the Board of Trade is, "to July 4th only, 45,452*l.*; and from June 4th only, entered by measure, 11,751 cubic feet."

ries.—Models of *Fruits* from Bombay.—Three large Cases are filled with the peculiar compressed cakes of *pharmaceutical vegetable products* of the U. S., prepared by a religious sect called "Shakers."

In the Room or large Gallery No. 3. is placed the beautiful series of *Woods* (against the wall) desiccated by Mr. Newton's patent process.

Room No. 4. *Two Cases* (47, 48.) contain an unnamed collection of *barks, woods, mats,* &c.

Room No. 5.—*Table-Case* contains numerous specimens of unnamed *fibres, drugs, fruits, oils, waxes,* &c.—*Chollet & Co.'s compressed Vegetables,* which have proved of great service to our brave armies in the Crimea, and can hardly be distinguished in flavour from fresh vegetables.

Room No. 6. (*in Table Cases.*) *Peats,* condensed without pressure, by Mr. Cobbold's patent process; these, having the character of coal, jet, &c., are capable of being turned into *inkstands, door-handles,* &c., and other preparations from peat. *Coal* and its products, and some *fossilised woods.*

Room No. 7. *Balsam-bog,* Bolax glebaria, already mentioned at p. 49.—*Poisoned arrows,* and *blowpipes* from the Amazon.—Numerous *drawings* of *vegetables* from Mr. Lawson's collection.

Room No. 8. (Ground-floor). On the window sills. *Case* containing vegetable products of the extreme arctic regions, brought home by successive expeditions sent in search of Sir John Franklin. —*Lattice-leaf,* Ouvirandra fenestralis. Madagascar.—*Wardian Case* with *living Ferns.*—Various large drawings are suspended from the Gallery.

Room No. 9. (a double room or passage at the foot of the stairs) contains a circular *plank* of a *wood* from the Island of *Ceram,* measuring 7½ feet in diameter. It is of an unknown tree, ordered to be cut by the late Captain Owen Stanley, R. N., for his father, the late Lord Bishop of Norwich.—Various samples of *foreign woods.*—*Five Cases* (91—96.) are filled with stems of *Tree-Ferns.*

The Staircase and Landing-place have numerous drawings (named) of useful and curious plants.—A portion of a comparatively *small* branch of the celebrated *Dragon's-blood Tree* of Orotava: a portrait of the Tree is suspended near it.—A noble *Tree-Fern stem,* Cyathea elegans, from Montserrat, reaches from the ground to the ceiling.

Room No. 10. (*Cases* 97. to 107. and *Table-Cases.*) Numerous specimens of unnamed woods from various countries.—The Ceylon collection of woods from S. Jayeteleke, Esq.—The beautiful *British Guiana* collection of *woods* was sent by H. Marriott, Esq.,

E. G. Boughton, Esq., and by the late Governor, Sir Henry Barkly, and first exhibited in the Crystal Palace for 1851.—*Spoons* and other *domestic utensils* made and used by the natives of Natal.—*Weapons* from Australia.—Examples of *roots* which fill to a great extent and choke underground pipes and tile drains : of agricultural interest.—The several kinds of woods employed for turnery purposes by Messrs. Holtzapfel.—One Case contains samples of wood cut through, showing the effects of injudicious pruning.—Several Cases are explanatory of the *destruction* to our timber and to grain by *Insects*, accompanied by specimens of the Insects, mostly collected and prepared by Professor Henslow, and others.—Other samples show the internal effect of *grafting*.—A Case of *Wasps'* and *Hornets' Nests*, may, at first sight, be considered rather Entomological than Botanical ; but the nests are curiously made by these little creatures of the fibres of *wood* masticated and converted into paper.—Adjacent, a Case is occupied by *materials* for *paper-making ;* with numerous varieties of *paper*, and *cable* from the " Royal George," together with *paper* made from it.—The Table-Case exhibits, among many other objects of interest, a series of the woods named, and samples exhibiting the mode of making *Tonbridge ware*, presented by Mr. Nye, the eminent manufacturer at Tonbridge ; and also a series of specimens showing the method of making the *Mauchlin Snuffboxes*, &c.

INDEX

TO SOME OF THE MORE INTERESTING OBJECTS.

PRINTED BY SPOTTISWOODE AND CO., NEW-STREET SQUARE, LONDON.